EUROPEAN WARFARE IN A GLOBAL CONTEXT, 1660–1815

European Warfare in a Global Context, 1660–1815 is a history of warfare, wars and the armed forces of Europe from the military revolution of the mid-seventeenth century to the Napoleonic era. Covering conflicts right through the period, Jeremy Black takes a revisionist approach to eighteenth-century warfare.

While Black discusses questions of military decisiveness and the military revolution, he is reluctant to stress changes in weaponry as the central narrative and the key analytical concept in analysing warfare in the period. He also questions common historiographical conventions, for instance the notion of *ancien régime* indecisiveness in Western warfare from 1660 until the French Revolutionary wars. Instead, his emphasis is on the importance of conflict in the period and the capacity for decisiveness in impact and development in method. Through this he extends the view beyond land to naval conflict.

Black also takes a global approach to this study of warfare. This approach is comparative, in the sense of considering Western developments alongside those elsewhere, but it also emphasises conflict between Western and non-Western powers. This approach reframes developments within the West, but also offers a shift in focus away from the standard narrative of warfare at this time, making *European Warfare in a Global Context, 1660–1815* an essential read for students of the period.

Jeremy Black is Professor of History at the University of Exeter. He is a leading military historian whose books include *Introduction to Global Military History* (Routledge, 2005), *Rethinking Military History* (Routledge, 2004), *The British Seaborne Empire* (2004) and *World War Two: A Military History* (Routledge, 2003).

Warfare and History
General Editor
Jeremy Black
Professor of History, University of Exeter

Air Power in the Age of Total War
John Buckley

The Armies of the Caliphs: Military and Society in the Early Islamic State
Hugh Kennedy

The Balkan Wars, 1912–13: Prelude to the First World War
Richard C. Hall

English Warfare, 1511–1642
Mark Charles Fissel

European and Native American Warfare, 1675–1815
Armstrong Starkey

European Warfare, 1660–1815
Jeremy Black

European Warfare, 1494–1660
Jeremy Black

The First Punic War
J. F. Lazenby

Frontiersmen: Warfare in Africa since 1950
Anthony Clayton

German Armies: War and German Politics, 1648–1806
Peter H. Wilson

The Great War 1914–18
Spencer C. Tucker

The Irish and British Wars, 1637–54: Triumph, Tragedy, and Failure
James Scott Wheeler

Israel's Wars, 1947–93
Ahron Bregman

The Korean War: No Victors, no Vanquished
Stanley Sandler

Medieval Chinese Warfare, 300–900
David A. Graff

Medieval Naval Warfare, 1000–1500
Susan Rose

Modern Chinese Warfare, 1795–1989
Bruce A. Elleman

Modern Insurgencies and Counter-Insurgencies: Guerrillas and their Opponents since 1750
Ian F. W. Beckett

Mughal Warfare: Imperial Frontiers and Highroads to Empire 1500–1700
Jos Gommans

Naval Warfare, 1815–1914
Lawrence Sondhaus

Ottoman Warfare, 1500–1700
Rhoads Murphey

The Peloponnesian War: A Military Study
J. F. Lazenby

*Samurai, Warfare and the State in
Early Medieval Japan*
Karl F. Friday

*Seapower and Naval Warfare,
1650–1830*
Richard Harding

The Soviet Military Experience
Roger R. Reese

Vietnam
Spencer C. Tucker

*The War for Independence and the
Transformation of American Society*
Harry M. Ward

*War and the State in Early Modern
Europe: Spain, the Dutch Republic
and Sweden as Fiscal-Military States,
1500–1660*
Jan Glete

*Warfare and Society in Europe,
1792–1914*
Geoffrey Wawro

*Warfare and Society in Europe, 1898
to the Present*
Michael S. Neiberg

Warfare at Sea, 1500–1650
Jan Glete

*Warfare in Atlantic Africa, 1500–
1800: Maritime Conflicts and the
Transformation of Europe*
John K. Thornton

*Warfare, State and Society in the
Byzantine World, 565–1204*
John Haldon

*War in the Early Modern World,
1450–1815*
Edited by Jeremy Black

*Wars of Imperial Conquest in Africa,
1830–1914*
Bruce Vandervort

*Western Warfare in the Age of the
Crusades, 1000–1300*
John France

*War and Society in Imperial Rome,
31 BC–AD 284*
Brian Campbell

*Warfare and Society in the
Barbarian West*
Guy Halsall

War in the Modern World since 1815
Edited by Jeremy Black

World War Two: A Military History
Jeremy Black

*War, Politics and Society in Early
Modern China, 900–1795*
Peter Lorge

*Warfare in the Ancient Near East, to
c. 1600 BC*
William J. Hamblin

*The Wars of the French Revolution
and Napoleon, 1792–1815*
Owen Connelly

*Indian Wars of Canada, Mexico and
the United States, 1812–1900*
Bruce Vandervort

*European Warfare in a Global
Context, 1660–1815*
Jeremy Black

First published 2007
by Routledge
2 Park Square, Milton Park, Abingdon, Oxon OX14 4RN

Simultaneously published in the USA and Canada
by Routledge
270 Madison Ave, New York NY 10016

Routledge is an imprint of the Taylor & Francis Group, an informa business

© 2007 Jeremy Black

Typeset in Bembo by Taylor & Francis Books
Printed and bound in Great Britain by
Antony Rowe Ltd, Chippenham, Wiltshire

All rights reserved. No part of this book may be reprinted or reproduced or utilised in any form or by any electronic, mechanical, or other means, now known or hereafter invented, including photocopying and recording, or in any information storage or retrieval system, without permission in writing from the publishers.

British Library Cataloguing in Publication Data
A catalogue record for this book is available from the British Library

Library of Congress Cataloging in Publication Data
Black, Jeremy.
European warfare in a global context, 1660–1815 / Jeremy Black.
p. cm. – (Warfare and history)
Includes bibliographical references and index.
1. Europe—History, Military—1648–1789. 2. Europe—History, Military—1789–1815. 3. Military art and science—Technological innovations. 4. War and society. I. Title.
D214.B565 2007
355.02'0940903—dc22   2006024745

ISBN10: 0-415-39472-4 (hbk)
ISBN10: 0-415-39475-9 (pbk)
ISBN10: 0-203-96482-9 (ebk)

ISBN13: 978-0-415-39472-7 (hbk)
ISBN13: 978-0-415-39475-8 (pbk)
ISBN13: 978-0-203-96482-8 (ebk)

EUROPEAN WARFARE IN A GLOBAL CONTEXT, 1660–1815

Jeremy Black

Routledge
Taylor & Francis Group
LONDON AND NEW YORK

FOR PETER WILSON

CONTENTS

Preface		*viii*
Abbreviations		*xi*
1	Introduction	1
2	Conflict between Westerners and non-Westerners	14
3	The nature of conflict	30
4	Warfare, 1660–88	48
5	Warfare, 1689–1721	65
6	Warfare, 1722–55	78
7	Warfare, 1756–74	93
8	Warfare, 1775–91	105
9	Warfare in the French Revolutionary and Napoleonic era, 1792–1815	118
10	Naval power	142
11	Social and political contexts	165
12	Conclusions	196
Notes		*204*
Select bibliography		*220*
Index		*223*

PREFACE

The development and impact of warfare in the Western or European world between 1660 and 1815 is a subject of great interest and importance, and this book is written in the conviction that it is most profitably examined in a global context. Within the serious constraints of the space available, this approach is a double one, first comparative, in the sense of considering Western developments alongside those elsewhere, as seen in the chapter on naval power, and, second, putting an emphasis on conflict between Western and non-Western powers, as in Chapter 2. This approach is not simply instructive from the point of reconsidering developments within the West, but also important because it offers a shift in emphasis from the standard narrative of the latter, with its abiding concern with conflict between Western powers; not that this is ignored in this book.

If the geography of this book is therefore unusual, the analytical framework is also not the customary one. There is a wariness about an emphasis on technology, and therefore a reluctance to stress changes in weaponry as at once the central narrative and the key analytical concept in explaining capability, success and change. There is also a questioning of the particular historiographical conventions for this period, especially the notion of *ancien régime* indecisiveness in Western warfare from 1660 until the French Revolutionary wars began in 1792. This is a convention which stems in large part from a perception of the military culture of the period which is misleading as far as the experience of conflict and the impact on civilians are concerned, points discussed in Chapters 3 and 11 respectively.

Furthermore, the period 1660–1791 is generally described in terms of falling between the putative military revolution discerned by Michael Roberts for the century 1560–1660, and the renewed burst of energy, change and significance seen as stemming from the French Revolutionary wars. The period 1660–1791 therefore takes on the character of pre-modern, more specifically the allegedly less consequential aspects of the early modern period, with its defining features apparently spotlighted by this understanding. In this study, the emphasis, instead, will be on the importance of conflict in the period, not least, but not only, the capacity for both decisiveness in impact and development in method. This is particularly clarified if attention is not restricted to land warfare in Europe between Western powers but, instead, extends to include naval and imperial capability and conflict, as well as struggles with non-Western powers. This re-reading of 1660–1791 puts subsequent developments during the French Revolutionary and Napoleonic period in a particular light.

A reconsideration of the chronological dimension can be related to a re-examination of the conventional idea of stages in a pattern of military change, with apparently ideal forms and paradigmatic powers defining military capability in a given period. This approach leads to an unfortunate focus on a small number of powers, with much of Europe ignored; or discussed largely in terms of these powers, for example, with reference to the idea of the diffusion of best practice, through the emulation of their methods. For this period, the powers that usually receive most attention are France, Prussia, Austria and Russia, and there is a general failure to appreciate the range and variety of warfare in Europe.

This book, in contrast, seeks to address this problem. It also offers a periodisation designed to encourage a re-examination of developments. For example, the decision to offer chapters on 1722–55 and 1756–74 is contrary to the usual focus on 1740–63 as a single period defined by the wars of Frederick II, the Great, of Prussia. The choice of dates is also designed to draw attention to the neglected significance of the wars of 1733–39.

I last tackled this period at this length and level in my *European Warfare 1660–1815* (London and New Haven, Conn. 1994). I was offered the opportunity of publishing a second edition with modest changes but, instead, have preferred to offer a new book that covers the same period but is extensively rewritten, entirely reworking the material in the light of more than a decade of research. The theoretical perspective of the 1994 book can also be replaced. There is now less need to discuss the questions of military revolution and decisiveness in the terms in which they were then framed, which in part reflected a historiography dating back to the 1950s; and, instead, a greater need to assess the issues of military cultures, both strategic and organisational. Cultural perspectives offer advantages without getting lost in some postmodernist linguistic turn. Furthermore, the linear assumptions of clear progress that underlay much earlier work are increasingly questionable.

The new book benefits from subsequent reflection and research, not least in considering the global context, since I wrote the earlier work in 1993. I have also profited from the opportunity to lecture on the period at many gatherings and institutions, and am happy to acknowledge the benefit of discussions with many other scholars. It was particularly useful to be invited to deliver the Ross Ellis Lectures in Military and Strategic Studies at the University of Calgary in 2006 and also to speak that year at the Norwegian Institute of Defence Studies, the symposium on France and the American Revolution at Brown University, the Naval War College, Yale University, Assumption College, and Trinity College, Dublin. Charles Esdaile, Jan Glete,

Richard Harding, Jürgen Luh, Mike Pavković, Rick Schneid, Manu Sehgal, Pat Speelman, David Trim, Sam Willis, Peter Wilson and Lorraine White made helpful comments on all or sections of this book in draft. It is a great pleasure to dedicate the book to Peter Wilson, a distinguished scholar of the period who is a friend of two decades standing and a very good fellow.

ABBREVIATIONS

Add.	Additional manuscripts
AE CP	Paris, Archives du Ministère des Relations Extérieures, Correspondance Politique
AST LM	Turin, Archivio di Stato, Lettere Ministri
BB	Bland Burges papers
BL	London, British Library
Bodl.	Oxford, Bodleian Library
CO	Colonial Office
CRO	County Record Office
Eg.	Egerton manuscripts
FO	Foreign Office
HL	San Marino, California, Huntington Library
IO	London, India Office Records
Lo.	Loudoun papers
NA	London, National Archives
SP	State Papers
WO	War Office

1

INTRODUCTION

War was central to the history of the period and to the experience of its people, while, in so far as long-term developments are concerned, these conflicts were also of great importance. As the result of the wars of the period, the Turks were pushed back into the Balkans, French hegemonic aspirations in Europe were defeated, and the fates of North America and India were settled, as was the struggle between Britain and France for maritime and colonial dominance. Furthermore, the Spanish empire was fatally weakened. Thus, the period was not only important to the rise of the West within the world, but also to the question of who was to dominate the West, which indeed answered the question 'which West?'. In particular, different versions of French predominance (Bourbon, Revolutionary and Napoleonic), in and beyond Europe, were offered and defeated during the period. Apart from these grand shifts – of territorial change, state-building and, more infrequently, ideological control – armies were also responsible for the maintenance of order and the defence of authority, whether against rioters, separatism, brigands, striking workers or religious dissidence.

The contrast, however, between different scholarly and popular perceptions of Western warfare in the period emerges clearly when the general account of this warfare on the world scale as dynamic and displaying greater capability than non-Western rivals, is set alongside the apparent situation within the West, which was allegedly characterised by stagnant or limited development under the *ancien régime* prior to the French Revolutionary wars, which began in 1792. This contrast reflects, in part, a difference in perception between the world scale, where the West appears as a transforming force, and, on the other hand, the standard view taken of conservative regimes within Europe, in this case those of the *ancien régime*. This view argues that conservative regimes are, simply through being such regimes, less able to adjust to the requirements of change, dramatically displayed by the French revolutionaries, and less willing to act decisively. In contrast, in this view, military capability requires the active embrace of change. This view plays a central role in the teleological, indeed Whiggish, assumptions that are so important to military history as an academic discipline. These assumptions, however, are flawed both in general and in specific cases.

INTRODUCTION

Much scholarship on military history is concerned with the search for dynamism in the shape of major change. The extent to which there was indeed change in portions of the period covered by this book, and the causes and nature of this change, have been debated, with particular attention devoted to the implications of the transition, at the close of the seventeenth century, from matchlock musket and pike (two separate weapons) to flintlock musket and bayonet (two capabilities in the same weapon that was wielded by one soldier); and to the use of column attacks by the forces of Revolutionary France in the 1790s. Furthermore, the respective merits of the developments seen under Louis XIV, Peter the Great, the Duke of Marlborough, Prince Eugene, Frederick the Great, and also in the closing years of the *ancien régime* in the 1770s–1780s have all been discussed. While notable, there were, however, no fundamental alterations in military capability or effectiveness, whether in force projection (the ability to deploy force at a distance) or combat during the *ancien régime*.[1] The absence of military revolution, or any equivalent to the sweeping organisational and technological transformation of the mid- to late nineteenth century, does not, however, imply that warfare in the meantime was static.

Whether discussing the period covered by this book, or more generally, arguments for military development have to be handled with care. Some of the literature is flawed because of its analytical approach, not least the questionable assumptions of the existence of a paradigm or model power setting the patterns for other powers, of a clear hierarchy in military capability and of an obvious tasking model for the military. Aside from these questionable assumptions, there is still the challenge of assessing the quality of change. In particular, the resilience of Austrian, Prussian and Russian forces in the 1790s, and the degree to which French war-making in the 1790s and 1800s built on pre-Revolutionary developments, indicate that the French Revolution did not make redundant what had come earlier, as is sometimes implied. This was a situation even more clearly exemplified by continued British success at sea before, during and after the Revolutionary period. This point suggests not only the strength of *ancien régime* structures, and the appropriateness of their tactics and operational and organisational characteristics, with reference to the parameters of the period, but also that the international politics of the 1790s bear a major responsibility for the failure to defeat Revolutionary France,[2] rather than simply military considerations.

Furthermore, once the notions that conservative societies lack the capacity for reform, and that they are necessarily weaker than revolutionary counterparts, are challenged, then the conceptual can be joined to the empirical in rejecting the notion of the *ancien régime* as redundant. This rejection also offers another approach to the issue of earlier change. Rather than seeing a fundamental problem of limited effectiveness in *ancien régime* warfare that required reform, it is possible to argue that the challenges were far more specific and contingent, being different for individual states at particular junctures. The focus should therefore be on considering the military tasks of the moment and those seen as likely to arise, which can be discussed in terms of what would subsequently be referred to as strategic culture, the phrase used to describe what were at the time apparently inherent strategic assumptions.[3] This may appear to offer a rather bitty and inconsequential account of the period, in place of the grand sweep of clear-cut and general developments in war-making, but, in fact, this approach accurately reflects the absence of a dominant teleology.

As recently as two decades ago, metanarratives of long-term military history were written in terms of such a teleology, specifically the move towards total war capability and doctrine then held to define modern warfare, especially with the maximisation of destructiveness through the enhancement of firepower. In such a context, the approach offered in the previous paragraph, with its focus on the particular requirements of the moment, would have appeared lame, and, in practice, as a recognition of the irrelevance of the *ancien régime*. Indeed, the period of this book was seen as important largely in terms of the progenitors of modernity supposedly offered by the American War of Independence (1775–83) and the French Revolutionary and Napoleonic wars (1792–1815). Now, however, as the multiple character of modern warfare can be better understood, so the very process of modernisation can be seen to involve far more continuity, and also non-linear development, than was hitherto appreciated.

In addition, modern interest in limited warfare makes aspects of the doctrine and practice of *ancien régime* conflict appear relevant; although, in fact, the limited character (within the parameters of what was judged possible) of this conflict should not be stressed. This was true for the experience of conflict, the degree of mobilisation of available resources (particularly manpower and money), and the impact on civilians, the last discussed in Chapter 11.

If change in the social politics and political consequences of force in early modern Europe can now be presented in more gradualist, and less revolutionary, terms, than would have been the case two decades ago, then this matches the long-term character of technological, scientific and intellectual trends. Once a chronological focus is added, then 'long term' can also mean slow, although that should not be seen as a criticism, not least because change was difficult and the character of military life and capability frequently intractable. Furthermore, in place of a 'big bang' seen in the language of military revolutions, or a triumphalist account towards clear improvement, comes the scholarly understanding that incremental change poses its own problems of assessing best practice, as well as the difficulties of determining whether it was appropriate to introduce new methods. Indeed, the definition and conceptualisation of novelty were themselves important issues of intellectual development. In practice, the use by modern scholars of models of diffusion, and of the language of adaptation, both make change in the past appear far less problematic than was the case.

INTRODUCTION

Naval power and warfare similarly demonstrate a variety also seen on land. This may appear surprising, as the focus on ships of the line in the literature appears to suggest a uniformity, while the period also saw the large-scale deployment of ships of the line in Mediterranean waters, for example by the English in the 1650s and from the 1690s, and by the Dutch and the French in the struggle over control of Sicily during the Dutch War of 1672–8, and the marked decline in the importance of galley conflict there. Furthermore, the decision by Peter the Great of Russia (r. 1689–1725) to build up a navy of ships of the line similar to that of other European powers rested in part on direct emulation, not least through Peter's visit to shipyards in England and the United Provinces (Netherlands) and through the hiring of Western experts. It was also maintained by his successors, most prominently Catherine the Great (r. 1762–96), under whom Russian naval power in 1769–70 was successfully deployed in the Mediterranean, albeit in a move dependent on British consent.

Nevertheless, there were other naval forces that also need consideration. In shallow waters, such as the Gulf of Finland, galleys proved more useful, and sailing warships were similarly of little value in rivers. River gunboats are generally ignored by naval (and military) historians, both of this period and of others, but they were important, particularly on the rivers of the Balkans and southern Russia, such as the Danube and the Don, in conflicts between the Turks and their neighbours. In 1702, the Swedish flotilla on Lake Lagoda was defeated by a far stronger Russian squadron.[4]

Even if the focus is solely on ships of the line, which were, indeed, the most important type of warship, it is necessary to note a parallel with land warfare in which the form, like that of musket–pike–cavalry cooperation in the late seventeenth century, and musket-plus-bayonet–cavalry cooperation over the following century, could include important variations; although, in assessing these, modern scholars have to address the contrast between contemporary doctrine and practice.[5] As with land forces, variations in naval tactics focused on distinctive specifications and objectives, particularly firepower, defensive strength, speed and manoeuvrability. These specifications, in turn, arose in large part in response to particular taskings. Thus, for warships, there were different emphases, including the extent to which ships were expected to cruise to the West Indies and to take part in line-of-battle artillery exchanges. Command skills on both land and sea involved understanding and taking advantage of the capabilities arising from specifications, and these are lost from sight if only a uniform account of weaponry and tactics is offered. It is also necessary to allow for variations in strategic culture at sea, as powers responded to the particular and changing needs and possibilities presented both by their situations and by the nature of their navies.[6]

An emphasis on variety in specifications and tasks can be extended to the equations of defensive strength and offensive firepower that helped determine the potency of fortifications and the success of sieges. Rather than assuming a perfect state of fortification, it is necessary to evaluate systems, not only with reference to the specifications that led to these equations, but also within the constraints of manpower and cost, each of which was also important in judging fitness for purpose. This returns attention from the theoretical and systemic to the particular, and, specifically, the political character of decision-making. Many positions were not strengthened in accordance with the latest fortification techniques, as they were too expensive to be comprehensively adopted. At the crudest level, the fortifications deemed necessary to withstand a major siege in western Europe were generally more than were necessary for most tasks in eastern Europe, and far more than was required to defeat a rebellion, although major fortresses in eastern Europe, such as the Turkish ones on the rivers athwart Russia's advance into the Balkans, such as Izmail and Silistria on the Danube, played a key role in campaigns.

It is also necessary to consider the particular requirements of colonial fortifications. For example, alongside the large sums spent by France in the 1720s–1730s on fortifications at Louisbourg on Cape Breton Island and on Pondicherry, its major base in India (where a rampart with bastions on the side facing the land was constructed in 1724–35, and a rampart on the coast in 1745), far less was spent by France on forts in the interior of North America and on the Louisiana coast. Indeed, in 1710, the woodwork at Fort Louis (Mobile) was so rotten with humidity and decay that the cannon could not be supported. However, these forts were not designed to resist artillery, unlike Louisbourg and Pondicherry. More generally, while fitness for purpose is a crucial concept when judging the applicability of weaponry and fortifications, such fitness is frequently misunderstood by putting the stress on the capacity for employing force, rather than the ends that are sought.

It is important also to assess the benefits that might stem if resources were employed for different purposes. Fortifications, for example, required large garrisons that could not readily be used for other purposes. Those near Susa designed to protect Piedmont from invasion from the Alps along the valley of the Dora Riparia were estimated in 1764 to require a garrison of 'near 4000 men' to defend them.[7] Such forces could act as an operational reserve, but, more commonly, garrison forces lacked mobility, flexibility and even combat effectiveness. More generally, when planning defences, it was necessary to consider the need for troops and cannon in different areas of possible attack.

This issue highlights the trade-offs and compromises involved in planning and prioritisation, and the latter underlines the role in military capability of the assumptions referred to as strategic culture. If a group maintains the same body of ideas about force and interests, and the same conditions of power and enemies, for long periods of time, it may define the same range of options and make similar choices among them, may create institutions of specific character and quality, may use them in distinct ways and may follow a strategic style, relying for example on attrition, such as commerce-raiding at sea, as against the search for an immediate knock-out blow. This approach is relevant for both land and sea power, and both in Europe and further afield, for example with the stress in French policy on the land rather than trade, on the army rather than navy, and on European, not transoceanic expansion. However, while the concept of strategic culture is clearly of value,[8] the detailed work required in order to flesh it out has not been tackled for most countries.[9] That helps explain the importance of assessing the social politics of command (the social groups that dominated military command and their attitudes), which was significant in the framing of strategic culture.

Linked to the limitations in modern scholarly knowledge of the subject is the popularity of the misleading concept of Zeitgeist or spirit of the age, as in the notion of an *ancien régime,* Enlightenment, or revolutionary Zeitgeist which explains the nature of war-making. In this approach, the move from one set of attitudes to another explains developments. In asserting such a coherence, often on the basis of only limited work, this approach, however, neglects this limitation and also underplays the diversity of attitudes and practices. As a result, the stress on the Zeitgeist is both empirically and conceptually questionable.

Concepts as varied as Zeitgeist, national interest and strategic culture indeed encourage reification (turning concepts into coherent causal factors) for reasons that can be actively misleading. Drawing on Neoplatonic assumptions about inherent reality, they assert a false coherence in order to provide clear building blocks for analytical purposes: they are more useful, in short, for assertion than as description. This point can be taken further by challenging the notion that the frequently used paradigm/diffusion models – the spread of the methods of an allegedly paradigm power – create, as is argued, either a cultural space in war-making or, indeed, bridge such spaces as this spread takes place. Instead, the emphasis can be on how the selective character of borrowing, both within, and between such spaces, reflected the limited validity of employing terms such as early modern, European or Western, as if they described an inherent reality or a widespread practice.

This criticism can be taken further by questioning the use of particular texts and contemporary or near-contemporary writers to describe the ideas dominant in particular conjunctures, and thus the military culture of the period. The works of Guibert, Clausewitz and Jomini have been particularly influential for this period. Aside from the problems faced in establishing the textual history of particular works and the details of their writers' intentions, there is the issue of the typicality of what survives, its influence, and the questionable nature of the relationship between literature and practice. Yet, it is valuable to consider texts, in order to provide alternatives to the use solely of military practice in order to understand contemporary assumptions. As the discussion of military culture focuses on perception – of norms, problems, opportunities, options, methods and success – this, however, poses more serious analytical problems than those of establishing the nature of battle, difficult as that is.

These points underline the looseness of the cultural description in military history, as well as the difficulty of employing the concept as either precise analytical term or methodology, but they do not undermine the value of the perspective. Instead, a synergy with the technological approach appears most attractive. Such a synergy would focus on issues such as the perception of improvement as well as processes of learning and norm-creation, both of which are important to the borrowing of ideas and practices. In this synergy, it will be necessary to be suggestive and descriptive, not assertive and prescriptive; but that is in accord with the nature of scholarship: history as question and questioning, not history as answer.

Similarly, strategic culture offers a possible synergy with political dimensions, as the pressures arising from international relations were important in the setting of goals and the tasking that helped determine doctrine and procurement and that drove strategy. In some cases, this process appeared clear. Joseph Spence, an Oxford professor who visited Turin in 1739–40, described 'the general interest and aims of this state' (the kingdom of Sardinia, which was based in Piedmont) as 'to keep always prepared for war; to weigh the strength of the greater powers that may fall out; and to join with that power, by whom they may get most in Italy, and whose increase of power can prejudice them least.'[10]

More generally, however, even if long-term goals seemed clear, the means to realise them were less so. There has not been a systematic study of strategic culture in the period, and indeed work on strategy is often limited: instead, the focus tends to be operational and tactical, or the war-and-society approach. The absence of such a strategic study, for example for French policy in 1715–55, suggests not only a clear research need, but also a dimension that needs to be built into other perspectives. In particular, it is unclear how best to address the role of *gloire,* loosely translated as the pursuit of glory, in military objectives. An emphasis on *gloire,* which is frequently seen as having been particularly important in the late seventeenth century, appears to offer a non-rational account of policy, one in keeping with the dominant role of rulers and the prevalence of dynastic considerations, but this emphasis, role and prevalence were certainly rational as far as the political purpose, ethos and structure of states were concerned. The acquisition of *gloire* brought important prestige to rulers and, in particular, helped strengthen their relationship with the aristocracy. That it was pursued so actively by hard-headed monarchs such as Louis XIV of France (r. 1643–1715) suggests that they saw purpose as well as pleasure in *gloire*.[11] A stress on the cultural context and goals of decision-making therefore emphasises the extent to which decisions to go to war went way beyond the material calculations of circumstances and state interests, the things that have featured in most standard accounts. The so-called 'new diplomatic history', with its emphasis on the culture of international relations, offers an instructive parallel to work on strategic culture.

The search for *gloire* was more widely diffused socially in the late eighteenth century, as a concern with public politics became more important, most obviously in Britain, but also in states that lacked powerful representative assemblies. There was a general shift from representational culture (directed both inwards to nobles and outwards to fellow monarchs) towards what might be called the political culture of national states. This shift involved a change of language in the way that strategy was discussed and legitimated. Furthermore, the declining importance attached to status and the growing awareness of material factors (and better measurement of them) affected strategy. The French Revolution also introduced a new ideological dimension. Under a new political dispensation in France, however, the concern with *gloire* was very much taken up by Napoleon (r. 1799–1815), who was assiduous in disseminating a triumphal account of his campaigning, as a major aspect of government propaganda. The contemporary memorialisation of success was important to the goals of warfare and to the way in which the experience of conflict was encoded into the peacetime norms of civil society.

If this approach provides the wider context for strategic culture, it still leaves more specific choices unclear. In particular, there is uncertainty about the impact in particular circumstances of the normative character of war and territorial expansion. These brought exemplary purpose and glory to rulers and aristocracy, and success acted as a lubricant of obedience in crown–elite relations. However, there were also many occasions on which international crises did not lead to war: for example, between Britain and Russia in 1720, the confrontation between the alliances of Hanover and Vienna in 1725–9, between Britain/France/Spain and Austria in 1730, between Portugal (backed by Britain) and Spain in 1735–6, between Britain and Spain/France in 1770 and 1790, or between Prussia and Austria in 1790 and Prussia and Russia in 1791; a far from complete list. The reasons for the avoidance of war require investigation from the perspectives of strategic culture and causes of war. Any military historian must also be prepared to explain peace.

Furthermore, the war plans drawn up in such crises also need investigation for the light they throw on political, strategic and operational assumptions, while the preparations provide indications about military capability and how it was perceived. Indeed, a systematic listing and analysis of such plans would be a worthwhile project. Plans also reflected the possibilities and problems created by alliances. In turn, the latter were both the products of strategic assumptions and yet also created issues of tasking and adaptation for these assumptions. The small change of diplomatic speculation related to the use of force to achieve strategic goals, including at a continental scale, as in 1739 when it was suggested that the French were arming Sweden and sending warships to the Baltic in order to threaten Russia and dissuade the latter from pursuing its offensive against France's ally Turkey.[12]

The emphasis in much of the scholarly literature on war and society, with the implication of a dualism of military and civil, offers the reflection that strategic culture can also be considered at different social levels or for various groups, particularly in states, such as Britain and the United Provinces (modern Netherlands), that had, in representative assemblies and a relatively

INTRODUCTION

free press, a means for such politics. This permitted the expression of different views. In the United Provinces, for example, there was support for the navy in the provinces of Holland and Zeeland and, conversely, for the army in the landward provinces.[13] Thus, strategic culture was not simply a matter of rulers, statesmen, generals and military theorists. Indeed, it was politicised, as in the United Provinces in 1684 where pressure from Amsterdam, encouraged by France, blocked William III's attempt to increase his army and paralysed his attempts to aid Spain, which ruled the neighbouring Spanish Netherlands (Belgium). Politicisation was frequently involved in the perception of danger, and this was made more contentious by the sense of flux that characterised international relations, a sense that is seriously underrated if the focus is on the schematic analysis of international relations and the allegedly structural interests of states. The unpredictabilities that stemmed from the role of monarchs were a particular cause of flux.

As far as the period of this book is concerned, the dualism of military and civil implied by war and society is unhelpful, as there was a considerable overlap. The extent to which 'the military' made a contribution to strategic cultures can be appreciated when it is realised that the military was not separate to political and social hierarchies, but intermixed with them, and, indeed, in part they were reflections of each other. This was not only true of 'absolutist' states, most obviously with the Table of Ranks introduced in Russia by Peter the Great, but also of those with powerful representative assemblies. For example, William III of Orange was the key figure in Dutch war-making in 1672–1702, while the British Captain General in the mid-1740s was none other than William, Duke of Cumberland, the second, and favourite, son of George II, and a major player in ministerial politics as well as military patronage.[14] Particular socio-political groups clearly had distinctive goals, but the extent to which these could be advanced through the overlapping worlds of court, ministerial, aristocratic and military factions, and the consequences for military planning, are unclear, and largely unstudied.

Yet, there were choices, most obviously when powers (i.e., rulers or ruling groups) engaged on several fronts and had to decide how best to decide between clashing commitments. Politics were clearly often involved. This was true for example of France in the Dutch, Nine Years', Spanish, Polish and Austrian Succession, and Seven Years' wars, as well as of successive Austrian Habsburg rulers driven to consider rival commitments in Italy, the Holy Roman Empire (essentially Germany and Austria), and the Balkans. The Emperor Charles VI faced, for example, the issue in 1717–18 of focusing on the conflict with the Turks and pushing Austrian territorial gains in the Balkans, or opposing expansionist Spanish schemes in Italy, a tension that recurred in 1737. Much of the discussion of the operational military history of these, and other, conflicts is limited as the result of a failure to relate it to the strategic dimension in this fashion. For Britain, it is possible to see a public debate, as with the tension between 'blue water' (maritime and transoceanic) policies and Continental interventionism, as in the War of the Spanish Succession and the French Revolutionary War, or with reference to sending troops to Germany in the Seven Years' War (1756–63).[15] Similar issues affected other states, but the discussion was less public.

Different strategic cultures also affected the perception of other states, and thus the dynamics of alliance diplomacy, both pre-war and wartime. This diplomacy, in turn, had an influence on policies and on particular military goals, underlining the extent to which there was a dynamic character to strategic culture. The difficulty of assessing this contributes, however, to an indeterminate nature for military history that is anathema to those who seek the certainties of uniforms, drill and (at least apparently) tactical detail; but this indeterminacy matches the uncertainty of contemporaries. Indeed, the clarity of many modern treatments of historical developments in warfare sits uncomfortably with, for example, the contemporary uncertainty about best practice.

This issue can be probed by considering the latter at the tactical and organisational levels. Action–reaction cycles, in which one power sought to counter the advances made by another, as when the French responded to John, 1st Duke of Marlborough's battlefield tactics in the 1700s, and the Austrians both to Frederick the Great of Prussia's oblique order attack in the 1750s, and to Napoleon's use of the corps system in the 1800s, indicated the dynamic nature of the challenges facing generals, but also the need in such a context to assess appropriate responses that were specific to particular problems, and thus the conditionality of best practice, a point that emerges even more clearly if conflict with non-Western forces is also considered. Seeking an advantage over the conventional tactics of his Austrian opponents, Frederick the Great in 1745 developed the attack in oblique order, so as to be able to concentrate overwhelming strength against a portion of the linear formation of the opposing army. Frederick devised a series of methods for strengthening one end of his line and attacking with it, while minimising the exposure of the weaker end. This depended on the speedy execution of complex manoeuvres for which well-drilled and well-disciplined troops were essential. By the late 1750s, however, the Austrians had developed effective counter-tactics, retaining reserves that could be moved to meet the Prussian attack. This proved an effective response to the oblique order attack (and indeed, when used by the French, to Marlborough's tactics), but that is not the same as defining it as best practice in contemporary Western warfare. Action–reaction responses to deficiencies revealed in conflict were seen in both Western and non-Western states. Just as Joseph II and Leopold II of Austria instructed the Nostitz Commission to examine problems and deficiencies indicated by the initial defeats at the hands of the Turks in 1788, so the weakness of the Turkish army seen at Russian hands in 1790–1 led to pressure for reform.

The challenges facing generals and admirals were also in part a matter of coping with established problems, particularly organisational and logistical issues, such as recruitment and supplies. The claim that logistical problems made the achievement of strategic objectives only rarely possible is not vindicated by the campaigning of the period, much of which delivered results in terms of victories and the seizure of territory, but these problems were serious.[16] It was also necessary for commanders to respond to new developments, but there was, moreover, the more general problem of gaining comparative advantage. The extent to which experienced and well-deployed infantry were frequently safe against frontal attack threatened a tactical impasse in infantry warfare that led to a particular need for skills in generalship, including, in the 1800s, the use of massed artillery to open the way for such attack. The naval parallel was even more apparent because of the difficulty of forcing a battle at sea if one side was determined to avoid it.

The dynamic nature of challenges was also true of the variety of environments that generals and their troops found themselves in. The large-scale British commitment in Spain during the War of the Spanish Succession (1702–13) was matched by the march of a large force to Blenheim in Bavaria in 1704, the furthest east commitment of such an British army on the European mainland, as well as the dispatch of an expeditionary force in an unsuccessful attack on French-held Quebec in 1711. Russian troops under Peter the Great moved into Estonia, Finland, Livonia and Poland in the early 1710s, and Mecklenburg (modern Germany) in 1716–17, but also into the Balkans in 1711 and along the Caspian Sea in 1722–3. More prosaically, French forces advanced towards Vienna and into Prague in 1741, although others were sent on an unsuccessful amphibious attempt to intervene against the Russians in Poland in 1734. Unexpected tasks included the Austrians facing civil insurrection in the city of Genoa in 1746, during the War of the Austrian Succession, as well as the French conquering Corsica in 1769–70, in the face of popular opposition in its difficult mountains. Yet again, the variety and unpredictability of military tasks, and thus the difficulty of judging capability and establishing a clear hierarchy of military proficiency, are underlined.

The variety of military tasking can be related to the nature of international relations. The rise of the sovereign state as a viable monopoliser of violence, a key development in western Europe by the 1660s and in eastern Europe by the 1720s, led in theory towards an international order based on sovereign equality, but, in practice, international relations seemed to remain within a single (still Christian) order, with a struggle for ranking along a hierarchical scale. The major shift after 1660 was the extension of this order to include non-Christians (which was achieved by the incorporation of the Turks from 1699), plus (by 1815) the final defeat of attempts to achieve the pole position in the hierarchy. The defeat of French hegemonic aspirations in 1812–15 confirmed a multi-polar system with five great powers (Russia, Prussia, Austria, Britain, France), several medium ones and a number of minor powers. Britain was the sole world power, but its power within Europe was limited.

The tension between research and theorisation is a common one in history, and military history is no exception. The statements frequently made about *ancien régime* warfare and the impact of the French Revolution, and the location of both in terms of general theories of military development, appear misleading, if not glib, as any consideration of the excellent detailed work available suggests. *Ancien régime* warfare within Europe in this period can be dismissed as rigid, anachronistic and limited in practice, and indecisive in result, only if a very narrow and misleading view of it is taken. At the same time, its dynamism and flexibility should not detract attention from the variety of military systems and conflict elsewhere in the world and the problems this posed for Western forces engaged with non-Western rivals. This comparative dimension is an important topic for discussion and should not be reduced to a consideration only of warfare between Europeans and non-Europeans, or, if the nature of both European-based settlement colonies in the New World and of Turkish rule in part of Europe is considered, Westerners and non-Westerners. Instead, it is instructive to assess struggles between non-Western forces as part of the analysis, not least because it helps explain the tasking of non-Western militaries and thus throws light on their response to Western militaries. This is a profitable field for future research.

Assessment of the wider global context further underlines the difficulty of considering best practice, and emphasises the diversity of Western warmaking. This assessment also contributes to the redundancy of the concept of indecisive or limited warfare, terms that would have meant little in the wider frame, most obviously at the very different scales of Indian territories conquered by British regulars, and Native American tribes, such as the Yamasee and Natchez, nearly wiped out by conflict with British or French colonial and regular forces respectively. This Western capacity drew in part, particularly in India, on the ability to recruit and train local native levies. How best to do so could exercise concern at the highest level – George III of Britain responding, in 1790, to the prospect of war with Spain and attacks on the Spanish West Indies (a war that did not break out), by suggesting that 'the measure of raising Blacks is much improved by attaching ten men to each company rather than forming a company of coloured men to be attached to each regiment.'[17]

At the same time, consideration of the large-scale Chinese expansionism in the period serves as a reminder that such successes should not be addressed simply in terms of Western capability, indeed a Western military revolution. Instead, the issue can be taken further, albeit at a smaller scale, if the expansionism of other powers is assessed, including, for example, Burma, Siam (modern Thailand) and, for the early nineteenth century, Egypt. Some examples, such as Persia (modern Iran) under Nadir Shah in the 1730s and 1740s, and Mysore in southern India in the 1760s–1790s under Haidar Ali and his son Tipu Sultan, can be dismissed as essentially short-term products of individual warrior-rulers that could not be sustained, and a contrast can then made with the more lasting and institutionalised nature of Western success (an approach conducive to theories of Western exceptionalism); but this argument

INTRODUCTION

needs to be handled with care because of the example of China. What might have been the short-term achievement of mid-seventeenth-century Manchu conquerors was, instead, grafted onto the continuities of the Chinese state, which helped ensure not only that major efforts were made, but also that the resulting gains were maintained.

The global dimension, which is probed in Chapter 2, thus invites questions about Western exceptionalism, narrowing it, in part, instead, on the naval power, considered in Chapter 9, and on the resulting capacity for transoceanic power projection. This dimension offers a more varied understanding of relative capability on land, and this poses questions about the applicability of the concept of an early modern Western military revolution. Within the West, this theory – of major change in *c.*1500–1660, followed by relative stasis in 1660–1791, until the onset of revolutionary warfare in the late eighteenth century – is also rendered questionable by an examination of the three chronological building blocks of the concept. Such a questioning does not mean, however, that the early modern Western warfare of the period was either unimportant or unchanging. As this book shows, it is not necessary for developments to be revolutionary in order for them to be significant.

2

CONFLICT BETWEEN WESTERNERS AND NON-WESTERNERS

The bells that rung out across Christendom to celebrate the total defeat of the Turks outside Vienna by German, Austrian and Polish forces in 1683, proclaimed relief, asserted the importance of the struggle between Christendom and Islam and truly marked one of the most overused of terms, a turning point. As such, this battle is an important aspect of another overused concept, an early modern military revolution, with specific reference to conflict between Westerners and non-Westerners. Although the general application of the concept is questionable, there is, however, no doubt of the importance of this warfare both for the societies directly involved and, indirectly, for states at a distance. Employing a geographical typology, this conflict can be divided between warfare around the margins of Western power in Eurasia and that involving Western forces at a distance from these margins. The first ranged from the vulnerable Western bases in north-west Africa, such as Tangier[1] and Oran, both of which were lost, through the Mediterranean, where Candia, the last Venetian position in Crete, finally fell to the Turks in 1669, along the land frontiers between the Ottoman (Turkish) empire and its Christian neighbours — Venice, Austria, Poland and Russia, and, albeit more marginally, between Russia and its Asian neighbours, particularly Persia and China.

The distance of the latter from the centres of Russian power made Russian military relations with China, however, more akin to the transoceanic power projection of western European maritime states. This power projection ranged across the world and can again be divided: between the expansion of already-settled colonies, some with their own military resources, and the establishment of new bases. The former was particularly true of the British, French, Spanish and Portuguese on the American mainland, and of the Portuguese in Angola and Mozambique in southern Africa; and the latter of the position in West Africa and, even more, Australia, but in some areas there was no stark divide.

In considering conflict, there is a tendency to assume a clear pattern of development in military capability predicated on an obvious best practice. The latter is generally seen for this period in terms of infantry volley fire, with the view that there was only one practice permitting the effective use of muskets in the field, and that the drill and discipline necessary to provide this became crucial to a distinctive Western way of war and ensured success against non-Western powers.[2] This argument presents those who did not adopt this obvious best practice in terms of failure of perception or execution. This is an analysis, however, that is questionable in itself, as it is normal to adopt practices that seem appropriate, and societies at the time generally have a better idea of this than scholars all-too-often writing with the condescension of posterity.

Furthermore, this analysis is empirically problematic. Work on the variety of military systems and methods seen in the West indicate that they were fit for purpose within the constraints of the relevant social systems, for example, light cavalry in Poland. This relationship between military methods and social systems had already been noted by writers of the period discussing, at least *en passant*, historical sociology, in the shape of the development of human society, most obviously Edward Gibbon and Adam Smith.[3] There is no reason why this perspective should not have been more generally applicable.

Military experience on the Western periphery, whether in the Balkans or in transoceanic colonies, existed in a dynamic relationship with practice within the West. Furthermore, the character of conflict between Western and non-Western powers varied, as did the success enjoyed by the Western powers. At times, victory was dramatic, and this was true not only in Europe, as at Vienna in 1683, but also further afield. On 12 February 1756, a British naval squadron under Rear Admiral Charles Watson demanded the surrender of Gheria, the stronghold on the west coast of India of the Angrias, a Maratha family, whose fleet was not only a factor in local politics, but had also been used for privateering attacks on Western trade in Indian waters (see p. 147). Watson's account of his victory emphasised firepower. In his description, when the Indians opened fire, he 'began such a fire upon them, as I believe they never before saw, and soon silenced their batteries, and the fire from their grabs [ships].' The five-hour bombardment set Tulaji Angria's fleet ablaze. Next day, the British warships closed in to bombard Gheria at pistol-shot distance, making a breach for storming, which led to its surrender. British casualties were slight: ten dead and seventeen wounded; and Watson reported, 'the hulls, masts and rigging of the ships are so little damaged, that if there was a necessity we should be able to proceed to sea in twenty-four hours.' Watson was supported by a body of troops under Robert Clive and by the Bombay marine (Bombay ships of the East India Company) under Commodore William James, who, in 1755, had similarly forced the surrender of Angria's bases at Severndaroog and Bankot. British confidence in their naval power had led Watson in 1755 to show his flagship, HMS *Kent,* to Mohammed Ali, the *nawab* (ruler) of the Carnatic in south-eastern India. British resources were readily available at Gheria, George Thomas noting 'We in [HMS] *Salisbury* fired 120 barrels of powder', and Thomas was certain of European superiority, 'A fine harbour . . . in the hands of Europeans might defy the force of Asia.'[4]

This sense was to be underlined two years later when Clive defeated Siraj-ud-daula, the young *nawab* of Bengal, at Plassey, a spectacular victory, with few casualties, by a small Anglo-Indian force using Western methods over a much larger army, albeit one wracked by dissidence. This victory, in which Clive lost about twenty-two men, laid the basis for the virtual control of Bengal and Bihar by the East India Company, the source of Britain's Indian empire, although that took several years of conflict. Plassey was followed by victories at Patna and Buxar in 1764, and in 1765 by the Treaty of Allahabad.

A sense of power was again shown in 1788 when Charles, 2nd Earl Cornwallis, the British Commander-in-Chief and Governor General in India (who had surrendered seven years earlier to the Americans and French at Yorktown), proposed to pursue Britain's claim to the *circar* (territory) of Guntur (Guntoor) against Nizam Ali of Hyderabad by force, using diplomacy simply to secure the settlement:

> It will be most expedient that our troops should march into the Circar on Captain Kennaway's arrival at Masulipatam [chief town]; and that our present Resident Meer Hussein should about ten days before inform the Nizam of our intention, giving the most positive assurances that our design was entirely limited to the taking possession of the Circar as our undoubted right by treaty.[5]

The result justified Cornwallis's optimism. Nevertheless, although British success in India, where the population far exceeded that of Britain, might seem a clear demonstration of Western military superiority, and in an important region of the world, yet a consideration of the episodes already cited, as well as of others, suggests that the situation was more complex. This provides a way to approach the subject of this chapter. To return to Gheria, Watson noted,

> The walls are very thick, and built with excellent cement, and the best stone I ever saw for such a purpose. We found upwards of two hundred guns here of different sizes, twenty three of which are brass, and six of them new field pieces with elevating screws, so that Angria was not without European friends, notwithstanding he was so common an enemy. There were also six brass mortars . . . and a sufficient quantity of ammunition of all kinds . . . had the garrison been provided with men of spirit and knowledge it must have been a much dearer purchase to us.

Watson also reported that Angria had been building a forty-gun warship. Although his emphasis was on the threat posed by the adoption of Western military technology, and therefore the danger posed by European powers willing to sell it, he also noted the strength of the fortifications. The same duality can be seen in subsequent conflict between Britain and Indian powers, ensuring that any approach to Indian capability largely in terms of the ability to absorb Western innovations is too narrow. At Plassey, Clive's success owed much to divisions among the Nawab's army.

Furthermore, the Indians already had long experience of using gunpowder weaponry. At Buxar, one British officer noted that the Indians kept 'up a very severe cannonade . . . in general their guns were well pointed', and again that Indian grapeshot was 'severe', did 'great execution' and stopped the advance of British *sepoys* (Indians trained by the British and in their service). Another officer commented on 'a very heavy cannonade' and wrote that the 150 cannon captured by the British were 'in good condition.' Both officers also noted the severity of the Indian cannonade at Patna (3 May 1764).[6] Yet, in each battle, the British were eventually successful. At Patna, British grapeshot halted the advance of hostile infantry and when a cavalry attack was launched by the Indians, 'a severe fire of artillery soon drove them back . . . we lost few men, but the enemy's loss was very great.'[7] At Buxar, Sir Hector Munro and 7,000 men of the East India Company's army, including 1,500 Europeans, defeated the 50,000 strong forces of the Nawab of Bengal, the Mughal Emperor, Alam II, and the Wazir of Awadh (Oudh). In this battle, the Indian army had more cannon, and used them to considerable effect, but British firepower was superior, and grapeshot and bayonets blocked the Indian cavalry. The Indians were driven from the field, although the battle lasted three hours and Munro's force lost heavily: 733 killed, wounded and missing, including sixty-nine Europeans.

Earlier, and as a salutary contrast, a significant attempt by another European agency to establish a strong position proved unsuccessful. The forces of the French East India Company, under the active Joseph-François Dupleix, Governor of Pondicherry, were successful in 1746 on the Adyar River, when *sepoys* defeated the cavalry and elephants of the *nawab* of the Carnatic, as well as in 1749 when intervening in a succession struggle in Hyderabad, and in 1751 when a French-Hyderabadi force defeated the Marathas; but it proved difficult to obtain any lasting victory over the Marathas, the lengthy campaign exhausted Hyderabadi finances, and Robert Clive undermined the French position in the Carnatic and helped install a British protégé. Dupleix's failure and his demands for men and money led to his recall in 1754.[8]

The limitations of Western techniques, tactical, operational and strategic, were, more generally, apparent, and British defeats or checks in India at the hands of Mysore and the Marathas from the late 1760s to the early 1780s were also instructive. These included the unsuccessful advance of a British army into the Ghats en route to Pune (Poona), leading to it being surrounded by the Marathas and forced to accept humiliating terms at Wadgaon in February 1779. Further south, in September 1780, a British force in the Carnatic was destroyed at Pollilore by Haidar Ali, ruler of Mysore. The British did not become the dominant force in India until 1799–1806, a period in which Mysore was finally overcome, with the capital falling and the ruler,

CONFLICT BETWEEN WESTERNERS AND NON-WESTERNERS

Tipu Sultan, being killed in its defence in 1799; while, with great difficulties, the Marathas were defeated in 1803–6. The future Arthur, 1st Duke of Wellington's victories at Assaye and Argaom in 1803 were crucial, but, in each, there was no clear-cut superiority for Western firepower. Instead, the Maratha artillery was effective in both battles, and British success, in what were very confused engagements, owed much to a willingness to advance at bayonet point.

Consideration of British setbacks and problems in India invites attention to Western difficulties elsewhere. Prominent failures were encountered at the hands of major states and also from less organised forces. In the first case, the Ottoman [Turkish] empire was, due to geography, the key opponent. Both Russia and Austria experienced failure as well as success in conflict with the Turks. The former included Prince Vasily Golitsyn's campaigns against the Crimean Tatars (allies and vassals of the Turkish Sultan) in 1687 and 1689, the Russian siege of Azov of 1695, Peter the Great's campaign against the Turks in 1711, and Austrian campaigns against the Turks in 1737–9 and 1788. Furthermore, British warships were hit hard in the Dardanelles by Turkish batteries in 1807. The most recent comparative consideration of Turkish warmaking argues that, due to various means of 'military acculturation', the Turks were able to match Western technological developments, so that, well into the first half of the eighteenth century, there were no major differences in gunpowder hardware. However, Western advantages are seen as stemming from major economic and administrative reforms in the late seventeenth century, which ended an earlier logistical inferiority, from better drill, command and control, and from the extent to which the Turks had to fight a number of opponents.[9]

The latter factor is instructive as it brings to the fore the prioritisation of tasks that is so important to the topic of strategic culture, as well as to the issues of appropriate force structure and doctrine. Furthermore, in contrast to Austro-Russian cooperation against the Turks in 1686–99, 1737–39 and 1788–91, the Turks lacked an ally in Europe. French diplomatic support and military advice was not the same as an alliance, either under the Bourbons or under Napoleon, while Sweden in 1738–9 and Britain and Prussia in 1791 proved unwilling to take enmity to Russia to the point of war. Moreover, bitter rivalry between the rulers of Turkey and Persia (Iran), which led to sustained conflict in the 1720s–1740s, ruled out alliance against Russia.

As well as failures against the Turks, there were also setbacks for Russia at the hands of Persia and China. In 1722, Peter the Great advanced along the Caspian Sea, his presence an indicator of the significance of the campaign. Derbent was occupied in 1722. The following year, Baku and Rasht were occupied and Shah Tahmasp of Persia was persuaded to yield the provinces along the southern and western shores of the Caspian. Rather than seeing this as a product of Western success and capability, however, it is necessary to note the peripheral character of Russian pressure in contrast to the chaos caused by the successful Afghan invasion of Persia, which culminated in the surrender of the Persian capital, Isfahan, in October 1722. A Persian revival under Nadir Shah and the loss of many soldiers to disease, indeed, led the Russians to abandon their gains by the Treaties of Rasht (1729, 1732) and Gence (1735). Further east, in 1685 and again in 1686, the Chinese successfully besieged the Russian fortress of Albazin and, in 1689, Russia signed the Treaty of Nerchinsk, acknowledging Chinese control of the Amur valley.

There were also setbacks at the hands of very different non-state forces. The Russians found it difficult to impose control among the Chukchi and Koryak tribes in the harsh environment of north-east Siberia. Further east, in Alaska in 1802, the Tlingits destroyed the recently founded Russian base of Mikhailovsk on Sitka. The Russians found the Tlingits tenacious foes, partly because they had acquired firearms from American and British traders and partly because, in this situation, the Russians derived little advantage from the naval power they had earlier used in spreading their power along the Aleutian Islands. In a very different terrain, the Russians also encountered problems in the north Caucasus, as indeed they also have done in the 1990s and 2000s, as much due to the intractability of their opponents as to the difficulty of the terrain. A Russian force was encircled and annihilated on the bank of the river Sunjt in 1785 by North Caucasian Muslims taking part in the holy war launched by Sheikh Mansur.

North America

In North America, European pressure was a factor throughout the period. This was true not only of the British, but also of the French despite their greater concern with power politics in Europe. In 1664–5, Louis XIV of France sent 1,200 troops who helped encourage the Iroquois, the dynamic Native Americans to the south of New France (the French colony centred on the St Lawrence valley), to terms, by the burning of villages and crops, which served to compensate for the impossibility of forcing the Iroquois to battle. This was a 'small war', but one that was different in type to that waged in western Europe (see p. 48–50), because the political context and the nature of the opposing military were both dissimilar.

As elsewhere with Westerners, forts were built in and near New France to consolidate gains and influence. In 1665–6, five forts were built along the Richelieu River, giving control over the route from the St Lawrence to Lake Champlain. Louis de Frontenac, the Governor, strengthened the French position on Lake Ontario by building Fort Frontenac in 1673, as well as Forts St Joseph (1679), Crèvecoeur (1680) and St Louis (1682), which consolidated France's presence between Lake Michigan and the Mississippi; but an attempt to crush the Western Iroquois in 1682 led to a humiliating climbdown after influenza and logistical problems weakened the French force.[10] Such problems were commonplace for Western units operating overseas when away from naval support.

CONFLICT BETWEEN WESTERNERS AND NON-WESTERNERS

An angry Louis XIV disgraced Frontenac in 1683, and an energetic new governor, Jacques-René de Denonville, arrived in 1685 with 500 troops, part of the 1,750 sent in 1683–8. As more generally, the numbers were far smaller than those deployed in European campaigns. Denonville successfully attacked the villages of the Seneca in 1687, and negotiated an acceptable peace with the Iroquois the following year. Further west, the network of French posts extended, with the foundation of Forts St Croix (1683), St Antoine (1686) and La Pointe (1693) south of Lake Superior, and Kaministiquia (1678), Népigon (1679) and La Tourette (1684) to the north. In 1688, however, the Karankawa wiped out the settlement at Fort St Louis founded by La Salle in 1682 on the Gulf of Mexico 400 miles west of the Mississippi delta. A strengthened French imperial drive could be seen at the close of the 1690s. Having devastated the Mohawk villages in 1693, and those of the Onondaga and Oneida in 1696, the French consolidated their position in the St Lawrence by a treaty with the Iroquois. Furthermore, the ability of Western powers to sustain initiatives was shown on the Gulf of Mexico, where, after the failure of the La Salle mission, a French base was founded in Biloxi Bay in 1699, the establishment of French positions at Mobile and New Orleans following in 1702 and in 1718 respectively.[11]

Nevertheless, imperial control, both over Native Americans and over non-national Europeans, could be precarious. The British found this in Nova Scotia in the 1710s–1750s, leading eventually to the expulsion of the Acadians, who were of French descent.[12] Furthermore, in some areas, Native Americans were able to fight 'European' Americans to a standstill almost to the end of the eighteenth century. This was even true near the eastern seaboard where, thanks to immigration, European pressure was concentrated: for example, the Yamasee nearly destroyed the British colonies in the Carolinas in 1715. The Cherokee resisted regular and colonial forces successfully in 1759–60, Governor William Bull of South Carolina reporting in 1761 that Anglo-American prisoners released by the Cherokee claimed that,

> their young men from their past observations express no very respectable opinion of our manner of fighting them, as, by our close order, we present a large object to their fire, and our platoons do little execution as the Indians are thinly scattered and concealed behind bushes or trees; though they acknowledged our troops thereby show they are not afraid, and that our numbers would be formidable in open ground, where they will never give us an opportunity of engaging them.[13]

Earlier, in a comment on the unsuccessful British-Creek campaign against the Spanish base of St Augustine in Florida in 1740, Lieutenant William Horton wrote, 'That the Indians are good to fight against Indians, and to waste the Spanish plantations, but not fit for entering breaches or trenches, or besieging a town regularly.' However, in Pontiac's War in the Ohio Valley in 1763–5, Native warriors not only demonstrated their military capability by driving the British back, but also the general adaptability of their way of war, specifically their ability to display new skills, including waging siege warfare and countering 'the actions of European armies in "regular battles."'[14] The British colonists (from 1776 independent Americans), however, had particular advantages in both spheres and also developed an effectiveness in small warfare, and a practice of devastating native civilians and agriculture, killing women and children as part of an effort to destroy native society. This hit native resources and made them less able to sustain conflict, as was seen in particular with the Sullivan expedition into Iroquois territory in 1779.[15] Earlier, a shortage of ammunition, and British scorched-earth tactics had led the Cherokee to opt for peace.

The role of religious considerations in the native struggle, for example Pontiac's understanding of Neolin, the Delaware prophet,[16] serves as a reminder that, like the struggle between Britain and the independent colonists in 1775–83, far more than resources were involved. It is important to consider the motivation of both sides when assessing their response to challenges and setbacks. The willingness to encounter major setbacks and yet fight on varied but was a key factor in explaining the intractability of some conflicts. Furthermore, such willingness could affect tactical, operational and strategic parameters.

Native strength was also apparent along the northern border of French expansion from Louisiana, as well as of Spanish expansion north from Mexico. The French military presence in Louisiana was weak,[17] and they suffered defeat at the hands of the Natchez in 1729 and the Chickasaws in 1736. In contrast, the isolated Natchez were defeated by joint French-Chickasaw action in 1731, and the Chickasaws intimidated by a larger French force into accepting a truce in 1739.[18] Moreover, the use of devastation – destroying villages and crops – as well as support from the Quapaws forced the Chickasaws to terms in 1752. Reliance on native support was typical of Western war-making outside Europe and a key aspect of the need to ground imperial ambitions in a management of local circumstances.

With the customary Western response of fortification, seen also, for example, with the Russians in Siberia, the Spaniards attempted to create an impregnable cordon of *presidios* (fortified bases) to anchor their interests and protect their territories. However, although Santa Fé resisted siege in 1680 in the Pueblo Rebellion, it then had to be abandoned, only being reoccupied in 1692. Furthermore, the Native Americans travelled between the *presidios* with no difficulty. The Apaches, Comanches and other Plains Indians were well mounted and armed, their firearms coming from trade with British merchants and with Spanish Louisiana, where there was a practice of trying to win them over by trade. The natives were also able to respond to Spanish tactics. Spanish expeditions, such as those against the Apaches in 1732 and 1775, were hindered by the fact that there were few fixed positions for them to attack, a habitual difficulty in asymmetrical warfare of this type. Furthermore, the Yuma Rebellion of 1789, in which Spanish positions were destroyed, thwarted plans of expansion through the Colorado Valley and into central Arizona. In contrast, the natives on the Pacific Coast lacked both guns and horses, so that Spain expanded her power in California rapidly in the 1770s, although, in 1775, the Ipais were able to burn the mission at San Diego. The ability to employ naval power gave the Spaniards a major advantage in California, not least in moving settlers and goods, and thus ensured a ready possibility for economic advantage from expansion, although California's coastal situation also ensured that other Western powers could challenge Spanish claims and establish positions, the Russians doing so at Fort Ross in 1812.

Near the eastern seaboard, the major rise in colonial population through immigration ensured that the natives were outnumbered, eventually heavily so. The combination of demographic and economic factors led Adam Smith to suggest that, although the Native Americans 'may plague them [European settlers] and hurt some of the back settlements, they could never injure the body of the people.'[19] However, there was not a population increase on a similar scale behind the expanding frontier of settlement in Spanish North America.[20] Due in large part to disease, large-scale immigration was not a successful option for Westerners in the tropics, but it was seen in more temperate lands, particularly Hungary and the Russian steppes.[21]

Western powers dominated the Caribbean, but in the West Indies, the resilience of non-Western forces was seen with the resistance of native Caribs in Grenada and St Vincent and with the ability of escaped slaves to resist Western forces, as on Jamaica in the 1730s. On St Vincent in 1772–3, the British encountered serious difficulties with disease, when operating against the Caribs. Slave risings in the New World were generally rapidly suppressed, for example the Stono Rising in South Carolina in 1739, the Revolt of the Tailors in Salvador in Brazil in 1798, as well as the Bahia Revolt of 1835. However, the slave rising on Saint-Domingue in 1791 was successful, despite major French efforts, and led to the establishment of Haiti as an independent state.

War in North America showed that the availability of Western weaponry (including gunpowder and shot), was regarded as a major advantage by native peoples, which made it important to limit the supply to hostile natives by rival Western powers. The adaptation to Western weaponry was not simply a matter of adopting it, but also of responding to the capabilities that such weaponry provided to opponents. Thus, the spear-wielding Australian Aborigines learned, tactically, to use the opportunities when their British opponents were reloading, and, operationally, to focus on attacking settler farms rather than to resist troops in combat. However, Aborigine numbers were small and affected by disease and disunity, while, relatively, there were numerous settlers, in what was a benign environment for them. As a result, Aborigine opposition was not sufficient to stop the British, who established their first base in Australia in 1788.²² Elsewhere, for example, as used by the Mura in central Amazonia, such methods could be more successful.

Furthermore, Western weaponry was used by natives in distinctive ways that reflected their own assessment of military practicalities, a process that included grafting this weaponry onto existing methods, which involved social as well as military constraints.²³ This was also true across the world, not only of peoples with small-scale state forms, such as the Maori of New Zealand, but also of major states, such as Burma.²⁴ Moreover, the use of Western weaponry by those who rebelled against imperial control, for example slaves in the West Indies in the 1790s and subject natives in Latin America, as in the large-scale, but unsuccessful, insurrection under Túpac Amaru in Peru in 1780–1, further indicated the variety of forms of military practice that were compatible with such weaponry.

As far as North America was concerned, the provision of weaponry was a key aspect of the geopolitical importance of Canada. If in hostile hands, it could lead to the supply of arms and ammunition to Native Americans resisting British, and later American, power in the future USA. Thus, the British capture of New France (Canada) in 1758–60 altered the position, lessening the options for Native American resistance, only for a rivalry between Canada and the USA to re-emerge with the War of American Independence, leading to unsuccessful American attempts at conquest in 1775–6 and 1812–14.

South America

There was also Spanish and Portuguese pressure on native peoples in South America, which is generally, but misleadingly, deemed not to have any military history between Francesco Pizarro's overthrow of the Incas of modern Peru in the 1530s and the successful wars of liberation against Spanish rule in the early nineteenth century. In fact, the conquistadores had taken only a fraction of South America and, over the following three centuries, conflict between Spanish and Portuguese colonies and natives continued, while there were also important, albeit unsuccessful, native rebellions, as in the Yucatán in 1746.

The Portuguese made considerable advances in Brazil, where the discovery of goldfields in the interior led to an intensification of activity and a native response. For example, resistance by the Paiaguá began in the 1720s, but they were affected by Portuguese expeditions, disease and attacks by the Guaicurú natives, and by the 1780s the Paiaguá had been largely wiped out. The use of local allies was important: Portuguese troops were unable to defeat the Caiapó, who ambushed Portuguese convoys, but the Bororo, under the leadership of a Portuguese woodsman, pressed them hard in a bitter war between 1745 and 1751. The ability to win and exploit local allies reflected the lack of native unity. This, and disease, all helped the Portuguese far more than any particular success in contact warfare. Portuguese firepower was important, but natives, such as the Mura in central Amazonia, learned to avoid it. The

Spaniards failed in the 1770s to subdue the Guajiros Indians in modern Colombia, and their control of the Mosquito Coast of Nicaragua was very limited.

Africa

In some areas, in contrast, there was no Western frontier of advance. For example, there were major setbacks for European powers in north-west Africa, a longstanding contact zone. Spanish-held Oran resisted attacks in 1667, 1672, 1675 and 1688, and Spanish-held Ceuta and Melilla, Moroccan sieges in 1720–7 and 1732, and 1774–5 respectively. The English, however, abandoned Tangier in 1683; the Spaniards lost Oran in 1708 (recapturing it in 1732, but evacuating it in 1792), as well as La Mamora in 1681, Larache in 1689 and Arzila in 1691; and the Portuguese lost Mazagam in 1765. More seriously, major Spanish attacks on Algiers were repelled. The Spaniards suffered heavy casualties in 1775, as their exposed troops were subjected to heavy fire, while their artillery could not be deployed speedily on the coastal sand. They were no more successful in 1784. It was not until the British bombarded Algiers in 1816 that it was defeated, and the city was not occupied until 1830, when the French seized it, the key 'tipping point' towards Western control in North Africa.

These Spanish efforts are a reminder of the mistake of thinking of conflict for the Western European powers simply in terms of warfare between such powers. Efforts against Algiers indeed were the foremost Spanish military commitment between the American Independence war and French Revolutionary Wars, while the 1732 Oran expedition had been a major effort. France made less of an effort in North Africa than Spain, but also encountered problems. The Bey of Tunis in 1741 seized the offshore island of Tabarca, which the French had purchased from the Lomellino family, defeated a counter-attack by a small French force in 1742 and sacked the French Africa Company's base at Cape Negre. A French attempt to land at Larache in Morocco in 1765 failed in the face of heavy fire.

Elsewhere in Africa, the attempt by the Portuguese to expand their influence up the River Zambezi into the kingdom of Mutapa (in modern Zimbabwe) was defeated by fever and by a widespread tribal rising in 1693–5: natives armed with spears were victorious, and the Portuguese were driven back to Tete on the lower Zambezi. The inaccuracy and low rate of fire of firearms lessened their value to the Portuguese in Angola and Mozambique, especially when confronting the dispersed formations common in southern Africa. Native forces were also able to take advantage of the vegetation cover and had their own missile weapons.

In West Africa, the Europeans were restricted to coastal trading bases and, while some survived attack by African forces, others did not, for example the French fort at Whydah captured by Dahomey forces in 1728. In the Senegal

Valley, gifts of guns were used to expand French influence, and in 1719 the French fortified Fort St Joseph to protect their trade along the river. They constructed riverboats for trade and protection. However, the French system of forts and boats depended on local cooperation. The usual Western position was one of dependence on the local African ruler, as part of a commercial relationship that included the purchase of slaves.[25] There was also African resistance to slavery, entailing both action against African slave hunters and against Westerners, the latter extending to risings on slave ships.[26]

Relative capability

The range of conflict between Western and non-Western forces makes it difficult to address the issue of relative capability and therefore the question of when a 'tipping point' to Western success occurred. This can be taken further if the need for more research is emphasised,[27] and if the variety of Western and, even more, non-Western, militaries, tasks and environments are considered. For example, there was no one manner of Indian conflict, and it was noteworthy that, alongside their proficiency with cavalry, Indian armies had also developed considerable effectiveness with artillery and infantry. There is, however, a misperception if Indian capability is seen largely in the perspective of how best to resist Western-trained forces. Instead, it is necessary to appreciate that, as was the case from the arrival of Western forces at the start of the sixteenth century, the major challenges came from other Indian forces:[28] in southern India, for example, from Mysore expansionism, while, in northern India, it was necessary to cope with attacks from foreign powers. In 1739, Nadir Shah of Persia defeated the Mughal army at Karnal and captured Delhi, while, in 1761, in another large-scale battle also greater in terms of combatants or casualties than Plassey, invading Afghans defeated the Marathas at the Third Battle of Panipat. The British East India Company infantry was more effective in the Lower Ganges Valley, in which it won key battles in 1757–64, and in operations on the Carnatic coast near their base at Madras, than in conflict against the Marathas and Mysore in regions that favoured their light cavalry.

It is also necessary to complement a focus on military operations, let alone fighting, with a consideration of other ways of increasing and employing power. Indeed, in part in responding to the temptation to search for single explanations, it is important not to emphasise conflict as opposed to more varied means of imposing will. Two examples from the New World illustrate this. First, the power balance in North America was shifted in the seventeenth century less by weaponry than by demography. The Europeans came to colonise, rather than to trade, and they, especially the British, came in increasing numbers. In contrast, hit by smallpox, native numbers did not grow. The impact of the Europeans' rising numbers was accentuated when advantages of mobility and logistical support enabled them to concentrate forces at points of contact. Second, in Central America, Nojpeten, capital of the Maya people known as Itzas, the last unconquered powerful native New World kingdom, fell to Spanish attack in 1697. In consolidating their position in the region, the Spaniards were helped by the rapid decline of the Itzas, in part due to epidemics, probably influenza and later smallpox, while a measure of control was imposed by moving the population into towns and Christian missions. Those who evaded the control lived in isolated forest areas, but were no longer able to challenge Spanish dominance. More generally, coercion and consensus both emerge as ways in which power was spread and sustained,[29] and they should not be seen as clearly separate in method and impact. This was also true of the non-Western world, with, for example, trade, disease and conflict, including the Shakushain War of 1669, all playing a part in the establishment and extension of Japanese rule over the Ainu of Hokkaido.[30]

The role of consensus and the extent of syncretism are important to the wider question of relative Western military effectiveness. So also is work, by Gabor Ágoston on the Turks and Randolph Cooper on the Marathas,[31] which questions conventional views about non-Western technological backwardness. The conventional emphasis on such backwardness is used to argue that mechanical philosophy in general, and Newtonian science, in the shape of scientific ballistics, in particular, were key 'tipping points' and definers of relative effectiveness.[32] However, whatever the accuracy of this approach by 1900, it was less clear, as far as land warfare is concerned, by 1815. In the eighteenth century, for example, the Indians did not have to adopt new technology but rather to adapt their existing usage and practices to meet the Western challenge. Acculturation, both in India and elsewhere, was therefore not simply a matter of copying. Nor was it solely a question of technology transfer and related models of organisation. Interaction could also be broader and involve more subjective issues, such as the readiness to use war, or to change practices, for example the treatment of prisoners.

India

The organisational potency of expanding and affluent Western powers emerges in the rethinking of British success in India, away from an emphasis on firepower, training, discipline and drill and, instead, toward the role of British credit, the expression of economic and fiscal strength and organisation, in the northern Indian military labour market. This credit meant that no single indigenous power would be able to claim the lion's share of South Asian military resources.[33] Furthermore, this financial strength helped the British create the military hybrid or synthesis important to their tactical and operational success, not least with the development of a cavalry able to provide mobility and to counter the consequences of that of their opponents, and with marked improvements in logistical capability.[34] More generally, the last was an important issue wherever Western forces operated. Their difficulties with logistics helped explain failure.

Indian rulers tried for the same process of synthesis, but less successfully due to financial, organisational and political factors.[35] These were both specific to military structures and more general. The former included the extent to which Indian armies relied on recruitment via semi-independent figures akin to the landed nobility of medieval Europe, whereas the East India Company relied on more direct recruitment and treated officers as a professional body subject to discipline.[36] British commanders and officers obeyed orders to a degree very different to that of Indian counterparts who were ready to change allegiances or to try to seize independent power. George Thomas, an Irish adventurer who, in 1797–1802, gained, by conquest, an independent principality north-west of Delhi, contrasted 'European troops, or Indian troops disciplined and conducted by European officers in which instance they may be considered as a machine actuated and animated by the voice of the commander', with the forces of the Rajah of Jaipur, 'in an irregular army where discipline never obtained, little can be expected from chiefs who in their actions are not stimulated by a sense of personal honour.'[37]

In part, such contrasts were a matter of tendencies rather than simple descriptions or abrupt contrasts, but these tendencies were nevertheless important. A more general contrast was the greater ability of the East India Company to deploy funds and credit, thanks to its oceanic trading position. In turn, its position in India enabled it to reap additional profit and thus to strengthen its financial credit.[38] The extent of this position in India helped ensure that British success was greater than the earlier achievements of other Western powers, although these had already indicated the advantages derived from commercial strength, not least in terms of eliciting local cooperation.[39]

Moving away from the emphasis on firepower in the case of explaining Western success is part of a wider process of reconsidering military capability. Parallels can be found with non-Western militaries. For example, as far as South Asia is concerned, the widespread dispersal of firearms ensured that a key to success was not their possession in itself, but rather organisation, not only in the case of armed forces but also in terms of exploiting agricultural resources and sustaining effective alliances.[40]

Financial and organisational strength were also seen in the global range of British power in the closing two decades of the period. This was exerted against Western rivals – France and its allies – but also in order to pursue British interests at the expense of non-Western forces. In 1788, the British established the first European colony in Australia and by 1815 this had been considerably expanded. So also had British interests in South Asia. Between 1806 and 1815, an expedition was sent into the Persian Gulf, war broke out with the Gurkhas of Nepal, and the kingdom of Kandy in Ceylon (Sri Lanka) was conquered. The latter two conflicts reflected the extent to which Western forces could campaign far from the ocean, although, as the British discovered in Egypt in 1807, operations inland could end in failure. Naval strength was crucial in projecting force and in maintaining the economic and financial links that were central to the articulation of maritime imperial systems. Thus, troops were sent to Louisiana from France to deal with the crisis created by the Natchez attack in 1729, and by the British from North America in 1772 to campaign against the Caribs of St Vincent.

Conclusions

Operationally, however, there had been a major shift away from a reliance simply on littoral activity and naval support, a shift seen clearly with the French and British in India from the mid-eighteenth century. To that extent, there was a degree of convergence with military activity on the West's borderlands in Eurasia. There the ability of Austria, and, even more, Russia to exert pressure reflected the extent to which Western activity was an aspect of far-reaching systems able to move resources. There were major difficulties, and, in the Russo-Turkish War of 1806–12, the Russian supply system broke down due to transport problems, but expedients were sufficient to permit a long-range Russian advance south of the Danube into Bulgaria.[41] Russian difficulties explain setbacks, but the latter did not prove the equality of the non-Western force. Instead, the Turks were not able to equal Russian war-making and fighting proficiency. The analysis of relative strength was more generally true for Western and non-Western forces by the early nineteenth century, although it was not invariably the case.

Furthermore, the Western presence was strengthened by the deliberate development of areas that were occupied, for example the steppe lands seized by Russia. Jean-Baptiste Colbert, who, in 1665, became French Controller General of Finances (and in 1669 Secretary of State for the Navy) wished to develop the St Lawrence Valley in New France as a source of food and industry that would complement the fishing off Newfoundland and the colonial goods from the West Indies, producing mutually supportive and profitable interactions in the French colonial world. In the 1660s and early 1670s, the government sent both money and settlers, including *filles du roi,* mostly orphans, whose immigration was subsidised in 1663–73 in order to offset the overwhelming preponderance of men and to ensure a larger population. French West and East India Companies were founded in 1664, and a large squadron was sent to the Indian Ocean in 1670, instructed to found fortified settlements near the Cape of Good Hope, on Sri Lanka, and in the East Indies, all areas with Dutch bases, at a time when France was planning for war.[42] It was not only this range of activity that was not matched elsewhere in the world, but also these aspirations.

To conclude on this point, however, would be to offer too clear-cut a portrayal of the initiative as lying with the Westerners and on their terms. This was true as far as deep-sea naval activity was concerned, but it is also necessary to appreciate the extent to which activity and aspirations were in the past moulded by the perception of non-Western activity. This was not simply a case of the discernment of opportunities, but also of anxieties about possible deteriorations in the situation for Western activity, whether commercial, political or territorial. Thus, for example, conflict between non-Western rulers was scrutinised carefully in order to assess what both activity and results meant. The Maratha victory over the Nizam of Hyderabad at Kharda (1795) was discussed by British commentators for its lessons for the British.[43] In part, this serves as a reminder of the importance of factoring in contemporary reactions to, and understandings of, military change when considering modern historiographical debates about interpretations of military change. More generally, such responses indicate the need to appreciate the range and dynamism of the two-way processes involved in the military interaction of West and East.

3

THE NATURE OF CONFLICT

In Europe, as elsewhere, battles and the smaller-scale clashes that were far more common were usually won by experienced and motivated troops whose dispositions had been well arranged. If forces were evenly matched, engagements were either inconclusive encounters, or were determined by other factors, such as terrain and the availability and, more significantly, employment of reserves.[1] In comparison, innovative ideas about deployment and tactics were not necessarily superior, while numbers alone were only of value if they were handled ably.

Technology

Technological change involved broader cultural, social and organisational issues. New weapons were developed: the socket bayonet and the flintlock musket in the late seventeenth century, the elevating screw for cannon in the eighteenth, as well as the introduction of conic ramrods, which allowed the reduction of the difference between the muzzle calibre and the ammunition calibre and thereby promoted more precise targeting. The rapid introduction of successful inventions or modifications in most Western armies, and the discussion of techniques such as drill and fortification, suggest that the importance of closing technological and other gaps was well recognised. There had to be a desire to respond, and technological change was therefore affected by cultural responses to innovation, just as the expectation of such change was important to developments in military capability and war-making.

Political and socio-economic contexts could play a role, not least in such developments. In England, the impact of the English Revolution during and after the English Civil Wars of 1642–6 and 1648 led to a major organisational transformation in both army and navy, albeit without any comparable shifts in weaponry.[2] Similarly, during the French Revolution, there was relatively little new science in French war-making, although it has been claimed that there was a new 'political will to break with tradition and apply scientific knowledge to the needs of the military on a scale that would have been difficult to imagine in the France of Louis XVI.'[3] Yet, apart from the short-lived corps of *aerostatiers*, it is difficult to think of conspicuous scientific innovations in the French military, and that is generally taken as the leading land power.

Alongside a perception of the essential sociocultural foundations of technological success, has come a view of technology as a direct 'social construct', with sociocultural forces shaping the technology. Once adopted, societies and cultural norms are themselves shaped by that technology, not least through the consequences of economic growth and change, but the underlying initial influences remain strong. From this perspective, warfare – the form and structure that it takes, and the technology it uses – emerges as a social construct. Dethroning technology from the central position in the narrative and explanation of military capability and change, does not, however, entail denying its importance. Instead, it is necessary to adopt a more nuanced approach to the different factors that play a role, considering them not as reified concepts that compete, but, rather, in a manner that allows for the multiple character of their interaction.

The role of research is clarifying the variety of factors that played a part in responding to, and shaping, suggestions for change. For example, recent work on French military engineering not only emphasises the importance of institutional conservatism, but also presents it as far from foolish. Marc-René, Marquis de Montalembert (1714–1800), who, from 1776 to 1797, advanced a series of bold fortification projects, was a man of great schemes but not one given to costing proposals or to detailed design. Furthermore, the practicalities of local topography were secondary in his thinking. What was important to Montalembert was the basic design, which determined whether or not a fortification was capable of withstanding attack. For him, reason was independent of nature and dominated it: the accidents of terrain and specificities of location could be subordinated to the theoretical plan. Montalembert's drawbacks help explain opposition from the French Corps of Military Engineers, but so also does the weight of the past, in the shape of the prestigious reputation of Sébastien Le Prestre de Vauban, who had been Commissioner General of the Fortifications under Louis XIV. Rather than see themselves as individual designers or inventors, as Montalembert did, engineers saw themselves as a corps, and this institutional response shaped the possibilities of change.[4] Hostility to Montalembert was in part due to his opposition to Vauban and his ideas. In contrast, the Prussians liked Montalembert and, in the nineteenth century, built their fortresses after his designs.

Similar work is necessary to understand other episodes. One major example is that of infantry weaponry. Towards the close of the seventeenth century, the development of the bayonet altered warfare in the West, although there was no comparable process on this scale elsewhere. The early plug bayonet, which was inserted in the musket barrel and therefore prevented firing, was replaced by ring and socket bayonets (attached to a metal ring around the barrel), adopted in France in 1689–92,[5] which allowed firing with the blade in place. This led in the 1700s to the phasing out of the pike, which was now redundant. Bayonets were a better complement to firearms in fulfilling the pike's defensive role, and also had an offensive capability against infantry and, on occasion, cavalry. This rapid change was largely carried out in the 1690s and early 1700s, and helped transform the shape of the battlefield. Earlier, it had been very complicated to coordinate pikemen and musketeers, in order to ensure the necessary balance of defensive protection and firepower, and this had led to a degree of tactical inflexibility, as well as to a density of formation that limited the possibilities of linear deployment over an extensive front. However, as a reminder of the danger of writing tactical history around innovations in weapons, linear formations had been developing prior to the new weaponry.

Tactics

The new tactical system encouraged the longer and thinner linear formations, as well as the shoulder-to-shoulder drill in order to maximise firepower, that were to characterise Western infantry in the eighteenth century, both within Europe and overseas. This ensured that, in battle, the tactical context was transformed, posing opportunities and problems for commanders and officers, not least, given the poor state of communications, those of coordination along a long front. An increase in the number of officers and non-commissioned officers helped address the problem;[6] although this could not compensate if, as often happened, there were problems with generalship, not least formation control and the difficulties of responding to the unpredictable flow of battle.[7] Deciding how best to employ and resist the enhanced firepower were also issues for commanders.

Similarly, changes in seventeenth-century Western naval warfare were responsible for the distinctive character of warfare at sea during the following century. Naval combat with artillery came clearly to replace the earlier preference for boarding. The development of line-ahead deployment and tactics for warships encouraged the maximisation of broadside power. This stress on cohesion reflected a move away from battle as a series of struggles between individual ships, although that element remained important because the nature of conflict at sea made it difficult to maintain cohesion once ships became closely engaged. Nevertheless, fighting instructions and line tactics sought to instil discipline and encouraged a new stage in organisational cohesion that permitted more effective firepower, one that was further enhanced when merchantmen ceased to appear in the line of battle of Western fleets (see p. 54).[8] Furthermore, the breakthrough in Western iron gun-foundry, which was not matched elsewhere, aided the production of large quantities of comparatively cheap and reliable iron guns, and this helped ensure the increase in the total firepower wielded by the leading Western armies and navies. The replacement of the wooden with the iron ramrod was another aspect of the dominant iron technology.

The bayonet–flintlock musket combination altered battlefield tactics, helping to lessen the role of cavalry, and ensured that casualty rates could be extremely high, particularly as a result of the exchange of fire at close quarters between lines of closely packed troops. Low muzzle velocity led to dreadful wounds, because the more slowly a projectile travels, the more damage it does as it bounces off bones and internal organs. Soldiers fired by volley, in a process designed to maximise the continuity of fire, rather than employing individually aimed shot. Despite the bayonets, hand-to-hand fighting on the battlefield was relatively uncommon, and most casualties were caused by shot. The accuracy of muskets was limited, which led to deployment at close range, while training stressed rapidity of fire, and thus drill and discipline. Within this pattern, there were variations in the order in which ranks (lines) or platoons fired, with the Dutch system of platoon fire becoming common during the War of the Spanish Succession (1701–14), except for the French. The Prussians then used a system with each platoon firing separately, producing a rolling fire, although the French continued to fire by ranks until the mid-eighteenth century. The speed of fire was enhanced by the use of paper cartridges, with the ball, powder and wadding in a single package. This made reloading easier and also allowed each soldier to carry more ammunition. Another instance of variety was the stress placed on the battlefield use of artillery either for counter-battery fire or for firing on opposing troops. The latter was a practice advocated by Frederick the Great of Prussia, and used by him, but was developed further by Napoleon who massed cannon successfully to that end.

Cavalry

On the battlefield, the infantry was flanked by cavalry units, but the proportion of cavalry in Western armies declined as a result of the heavier emphasis on firepower as well as the greater per-capita cost of cavalry. There were also particular supply problems for cavalry, with the need for large quantities of fodder, which was bulky to transport. Cavalry was principally used on the battlefield to fight cavalry: advances against unbroken infantry were uncommon, and at Minden (1759) unsuccessful, although infantry, unless it could rapidly reform, was vulnerable to attack in flank and rear. Cavalry played a crucial role in some battles, such as the British victory over the French at Blenheim (1704) and the Prussian over the French at Rossbach (1757), and cavalry–infantry coordination, or at least combination, could be important. At Fraustadt (1706), a Swedish army defeated a Saxon force twice its size, the numerous Swedish cavalry enveloping both Saxon flanks, while the relatively small Swedish infantry force held off attacks in the centre. Much depended on the terrain and cover. Due to the many hedges on the battlefield, Roucoux (1746) was very much an infantry battle. In general, however, cavalry was less important than it had been in the past. The degree to which the social prestige of being an officer in cavalry units nevertheless remained high reflected the extent to which values did not necessarily reflect practicalities. Despite this, the infantry was the key arm.

Firepower

Unbroken infantry was more vulnerable to artillery than it was to cavalry, especially because of the close-packed and static or slow-moving formations that were adopted in order to maintain discipline and firepower. The use of artillery increased considerably during the century, and, by 1762, Frederick the Great, who had not, initially, favoured the large-scale use of artillery, was employing massed batteries of guns. Cannon became more mobile and standardised. The leaders in this field were, in the 1750s, the Austrians, who had made their field pieces more mobile by reducing their weight, and, from the late 1760s, the French, under Jean-Baptiste Gribeauval, were the leaders in this field. The greater standardisation of artillery pieces led to more regular fire and thus encouraged the development of artillery tactics away from the largely desultory and random bombardments of the seventeenth century in favour of more efficient exchanges of concentrated and sustained fire. Artillery fire therefore acquired the key characteristics of its infantry counterpart. Artillery was employed on the battlefield both to silence opposing guns and, more effectively, in order to weaken infantry and cavalry units. Grape and canister shot were particularly deadly: they consisted of a bag or tin with small balls inside, which scattered as a result of the charge, causing considerable numbers of casualties at short range.

The notion of ready improvement, however, has to be qualified by an appreciation of the difficult trade-off in field artillery between mobility and weight. Heavier field pieces provided more effective fire support, but that, in part, depended on the battle being relatively static, so that mobility was not at a premium. Effective artillery was also important to the overseas capability of Western forces on land and at sea. In India, the British East India Company steadily increased the number of cannon in its field forces.[9] The development in the mid-eighteenth century of a new system of casting cannon – casting them solid and then drilling them out, rather than casting a hollow barrel – improved effectiveness by decreasing windage (the gap between the barrel and the shot). This made it possible to use smaller charges and thus to lighten barrels and increase mobility.

There was a tension between firepower and shock tactics, for infantry, cavalry and siegecraft. The different choices made reflected both the views of particular generals, and also wider assumptions in military society; they were not dictated by technology.[10] An emphasis on the attack can be frequently seen in the case of Gaelic, Polish, Russian and Swedish forces. In Gaelic warfare in Highland Scotland, there was a stress on the tactical offensive by infantry,[11] although neither Gaelic nor eastern European war-making should be simplified, and, in particular, it would be as foolish to have an ideal account of eastern European warfare, as it is to have one of western European, or indeed Western (European excluding Turkish) warfare as a whole. War-fighting and the trajectory of military development varied, as can be seen by contrasting Austria, Poland and Russia, let alone Turkey: for example, the Polish emphasis on light cavalry was not matched by a stress on infantry equivalent to that of Austria or Russia.

Marshal Saxe, the leading French general of the 1740s, and Frederick II, the Great, of Prussia were also skilled in attack. Saxe's victory at Fontenoy (1745) was a defensive one, but it was followed by offensive victories at Roucoux (1746) and Lawfeldt (1747). In short, the strategic, operational and tactical offensive had not been banished by the enhancement of firepower from the 1690s, and this prefigured the situation in the twentieth century. Shock also won over firepower in the case of cavalry tactics. Furthermore, shock played a major role in victories over non-Westerners, particularly in the defeat of the Turks by the Austrians at Vienna (1683), Zenta (1697) and Belgrade (1717), and by the Russians in 1770 and 1774.

Firepower also played a major role in sieges, but many fortresses, often major ones, fell to assault, such as Prague in 1741 and Bergen-op-Zoom in 1747 (both to the French), and Izmail in 1790 to the Russians, although the storming of major positions was generally preceded by bombardment and the deliberate preparation this entailed with the digging of trenches and the construction of artillery platforms. The role of storming indicates the need not to treat battle and siege as necessarily contrasting, for storming attempts could lead to large-scale clashes in the breaches made in walls. This can be taken further if it is noted how many battles arose as a result of relief attempts on behalf of besieged fortresses (for example the Portuguese victory of Montes Claro in 1665, Vienna in 1683, and Fontenoy), or followed soon after the fall of a position, as in 1662 when an Anglo-Portuguese army defeated the Spaniards outside Evora, soon after regaining the city. The storming of lesser positions, in contrast, was not generally preceded by bombardment. Instead, it was frequently an aspect of the rapid moves of 'small war' (see pp. 48–50), with surprise attacks launched without the delays or concentration of force required if artillery was to be deployed.

An emphasis on shock does not necessarily lessen the value of firepower, especially because the latter could be used to prepare for the assault. However, it suggests that any account that approaches tactical success in terms of the technological (weaponry) and organisational (tactics, drill, discipline) factors that maximised firepower is a limited one. The real point of drill and discipline was defensive: to prepare a unit to remain intact in the face of death, regardless of casualties. The issue of shock versus firepower was not as important as a unit remaining able to act, and tractable to its commander, while receiving casualties, which could be very heavy.

Far from being static, wars saw a tactical and operational responsiveness to circumstances, not least the war-making of opponents. Thus, during the Seven

Years' War (1756–63), as everyone sought to avoid the mistakes of the previous year's campaigning season, warfare was shaped by the fluid dynamics of the contending armies. Initially, Frederick the Great relied on cold steel. At the battle of Prague (1757), for example, he planned to roll up the Austrian position to the east of Prague from its right, but the Austrians were able to move much of their army to cover this flank. The Prussians were therefore forced to make a frontal attack. Advancing with shouldered muskets, the Prussian infantry was repulsed with heavy losses from Austrian cannon and musket fire, and with Field Marshal Schwerin killed. However, the Prussians were then able to advance in the gap between the Austrian main army, still facing north, and the units that had defeated Schwerin. The exposed wing of the latter was attacked, while, at the same time, the Prussian cavalry defeated its Austrian opponents thanks to a flank attack. The principal Austrian position was rolled up from its right flank and, when the Austrians rallied, they had to withdraw due to a Prussian threat to both flanks. Nevertheless, although the Austrians lost about 14,000 men, including 5,000 prisoners, the Prussians had 14,287 dead and wounded.

Thereafter, Frederick was to place more emphasis on the tactics of firepower, for example at the Battle of Leuthen later that year. He also developed artillery-based tactics, which were not simply a response to the capability of cannon, but also a reaction to the problem posed by the successful use of hilly positions by the Austrian Field Marshal Daun. Furthermore, Frederick had to address the deficiencies of Prussian offensive tactics in the face of Austrian and Russian defensive positions, as at the Battles of Kunersdorf and Torgau. Once an attack had failed, it was difficult for the Prussians to regain the initiative. Organisational factors also played a role in tactical choices. For example, as the Prussians, afraid of desertion, did not like to abandon dense formations in favour of loose counterparts (which would not anyway be able readily to provide concentrated volley fire), there was, instead, in 1762, a use of dispersed columns in attack, as at Burkersdorf and Freiburg.

During the Seven Years' War, a rapid learning curve could also be seen with the British in North America, as they adapted to the tactical and operational exigencies of conflict in the interior with the French, and, soon after, in Pontiac's War (1763–5), the Native Americans.[12] Again, this indicated the variety and flexibility of *ancien régime* warfare, although this was not such that all differences with non-Western military practices were elided, or even lessened.

The nature of conflict

There is often an emphasis on conventions of *ancien régime* fighting that suggest that it was restrained and limited. Among officers of the period, there was indeed a pseudo-chivalry that reflected their sense of being members of a common profession and, in large part, aristocratic caste, and also conventions of proper conduct. In 1758, Lord George Sackville (who was later responsible for British strategy during the American War of Independence) reported of the opposed British and French forces in Germany, 'Our sentinels and advanced posts are in perfect harmony and good humour with each other, they converse frequently together and have not yet fired though officers go out of curiosity much nearer than there is any occasion for.'

The following year, Captain William Fawcett noted that the advanced posts of the two forces were very close,

> which afford a sight extraordinary enough, to those who are not acquainted with the formalities of war. The French, to do them justice, are a very generous enemy and above taking little advantages: I myself am an instance of it, amongst many that happen almost daily: Being out a coursing a few days ago, I was galloping at full speed after a hare ... into a thicket, where they had a post of infantry, and must infallibly have been taken prisoner, if the officer commanding it had not showed himself; and very genteelly called out to stop me. We frequently discourse together.[13]

Such episodes have tended to characterise the image of pre-French revolutionary warfare, with particular attention devoted to the British and French at the Battle of Fontenoy (1745) and a willingness there to let the other fire first. Indeed, the treatment of prisoners improved from the late seventeenth century: in the eighteenth century, the exchange of prisoners on a large scale became common, while the rituals of surrender remained important.[14]

Nevertheless, indications of restraint, such as the relative rarity of allowing the plunder of stormed positions,[15] are less than the full picture. Aside from the actual horrors of the fighting, including the use of case-shot against close-packed rows of soldiers, or the slaughter of fleeing infantry by cavalry, and the primitive nature of medical care, soldiers were often brutal in their treatment of each other and of civilians. After Dettingen (1743), British soldiers plundered and stripped their dead and wounded compatriots.[16] At the Katzbach (1813), the victorious Prussians did not take French prisoners. The horror of war was captured by Richard Browne, who served in the victorious British army at Minden (1759), writing to his father,

> I thought formerly I could easily form an idea of a battle from the accounts I heard from others, but I find everything short of the horrid sense and it seems almost incredible that any can escape the incessant fire and terrible hissings of bullets of all sizes, the field of battle after is melancholy, four or five miles of plain covered with human bodies dead and dying, miserably butchered dead horses, broken wheels and carriages, and arms of all kind ... in the morning on the ground in our tents was pools of blood and pieces of brain.[17]

The commonplace and coarsening nature of killing was noted by Fawcett the following year,

> the destruction of two or three hundred poor wretches, is looked upon as a mere trifle here, wherever there is any point to be carried, which is thought of consequence . . . 500 or 1,000 fine fellows, in full bloom and vigour ordered to march up, to possess themselves of an eminence, an old house, or windmill, or other particular piece of ground, with a certainty of one half of them at least, being, at the same time, exposed to certain death, in the doing of it. Nevertheless, this does, and must happen almost daily, so long as the war lasts.[18]

Supplies

The brutality and harshness of conditions were also a product of repeated problems with logistics. The combination of a military system that placed a premium on numbers, both on land and at sea, a slow, cumbersome and labour/animal/ship-intensive supply system and economies yielding only a modest surplus created serious logistical problems. Generals relied both on the supplies they would obtain, by whatever method, from the regions in which they campaigned, and on those obtained from their home bases. This was a system of shifts and expedients, and, when it broke down, the troops suffered. The size of forces posed formidable logistical problems, especially once conflict had lasted for longer than about two years, exhausting both stored supplies and those extorted from the areas of campaign. The extent to which campaigning recurred in the same areas limited the value of contribution systems, although they were employed extensively. Examples included by the Austrians, for example, after they conquered Bavaria in the Spanish and Austrian Succession Wars in 1704 and 1742 respectively, and by the Russians repeatedly in Poland. The alternative of purchasing supplies in the field was similarly made difficult by the limited agricultural surpluses of many areas of Europe, especially after a number of campaigns. The weakness of government credit, especially in wartime, was also a factor. When units had to rely on seizing local supplies and on forced loans, they proved a heavy burden to the population and antagonised them.[19]

Despite serving Britain, the most creditworthy of the major states, John, 2nd Duke of Argyll was obliged to report in 1711 from Barcelona, where his forces (taking part in the civil-war dimension of the War of the Spanish Succession) were short of powder and cannon,

> having with greater difficulty than can be expressed found credit to keep the starving in their quarters . . . which I do not see how we shall be able to do any longer, for the not paying the bills that were drawn from here the last year has entirely destroyed her Majesty's

credit in this place . . . besides that the contractors for the mules to draw the artillery and ammunition and carry the bread will by no means be persuaded to serve any more till we have money to pay them.[20]

Campaigning in the Low Countries, one of the more favourable regions for supplies, and, like Barcelona, an allied area, Sir John Ligonier of the British army wrote in 1746 of 'the want of all things both horse and wagons for our [supply] train . . . we have had four days long marches over a country less plentiful than the Highlands [of Scotland], and yesterday and today the soldiers are without wood or straw and have extreme bad water.' Two years later, Edmund Martin reported from near Eindhoven in the allied United Provinces about fevers and deaths in the British forces, 'the wetness of this country, the bad stagnated ditch water we drink, the bad food . . . we lie in barns and open cowhouses with little or no straw . . . The Hollanders in their camp and garrisons are the same; I hear the French are too.'[21] Nevertheless, the Low Countries, like Lombardy, was relatively well provided for thanks to its productive agriculture. Alongside the political and strategic importance of these regions, this helped account for the tendency to campaign there.

Problems with supplies also occurred elsewhere. Campaigning in Westphalia in western Germany in October 1760, Fawcett revealed his sympathies, 'the poor soldiers lying upon the damp, and without straw, occasions great sickness in the army.' A year later, Major-General George Townshend offered a far from heroic account of operations in the region: 'this exhausted, pillaged, infected country where nothing but chicanes in war, and misery and despair to the inhabitants remains. Our army is really a scene of indiscipline, weakness and almost despondency. I never saw so much pillage and desertion; it is general.'[22]

Such comments both cast light on the commonplace frictions of war and also on the situation a century after the end of the military revolution described for the period 1560–1660 by Michael Roberts. British examples are particularly instructive, because commercial and financial strength and organisation are supposed to have made Britain an effective military state,[23] and certainly did so in comparison with other states. Yet, the British encountered major difficulties, and it is useful to cite the example of the expeditionary force sent to Portugal in 1762. This neutral ally of Britain appeared to France and Spain to be a vulnerable target that could be overrun, hitting British trade, and also exchanged in a general peace treaty for British colonial gains during the Seven Years' War. Portugal was certainly weak, as it had also been when the British earlier sent military support.[24] A state of the Portuguese army presented by its government to that of Britain in 1761 revealed that, although, in theory, the army was 31,000 strong, in fact it only had 16,500 men,[25] and was short of artillery, horses and supplies. Spanish successes in overrunning weak and poorly defended Portuguese fortresses in early 1762 led to urgent requests for British reinforcements, but these were delayed in the English Channel by contrary winds in late June. The following month, Brigadier Frederick provided a depressing account of the logistical problems he faced on the march to Santarem, which, in part, arose from the poverty of the region. Arriving at Porto de Mugen, Frederick had found no beef or bread prepared for his troops and it proved impossible to obtain adequate supplies,

> all the bread that the Magistrate said he could possibly get before they marched was two hundred small loaves which was so small a quantity it was impossible to divide amongst the men. I ordered the regiment to march the next morning at half an hour past three, but the carriages for the baggage not coming at the proper time it was past six before they began their march. It was late in the day before they got to Santarem when Colonel Biddulph reported to me that by the excessive heat and sandy roads that above half the regiment had dropped behind and was afraid many of the men would die, on which surgeons were sent to their assistance . . . the magistrates had provided no quarters for them, neither was there beef or bread for the men . . . they were fainting with the heat and want of food.

Clearly the pressure on the soldiers was acute, 'nine men of the Buffs died on the march yesterday.'[26] Frederick's letter lends vivid point to the complaints of British generals over Portuguese supplies, especially of horses, mules, bread, forage and firewood. However, the Spanish failure in 1762 to exploit their early successes by a march to capture Oporto, the major town in northern Portugal, proved operationally decisive, especially in the context of increasingly successful Anglo-French negotiations designed to bring the war to a close. This failure also reflected a more general problem for many armies and navies, at least the British navy, that of attaining tactical and operational effectiveness in the opening campaign of any war.

Nevertheless, the conflict continued, and the correspondence of British officers throws considerable light on the problems they encountered, especially with communications and in cooperating with the Portuguese. Captain Fraser Folliott reported to the British commander, John, 4th Earl of Loudoun, 'I have examined the ford at Belvere over the Tagus situated at Ortiga, and do find five feet water almost all the way over, on the bottom are large rocks and stones, and the current very rapid – I have also been on the other side of the river and find the roads everywhere impassable for wheel carriages.' And later, 'near a third of my company being sick most of them of fevers, if a surgeon's mate could be spared for four or five days, he would be of infinite use.' Folliott wrote to Wilhelm, Count of Schaumburg-Lippe, the minor German prince who had become commander of the Portuguese army (and who wrote treatises on the art of war), 'It will be very difficult for me to carry that part of your Highness' commands into execution relating to gaining intelligence, without a person who understands the Portuguese and English languages. I have been much distressed already for such a person, and four or five times have I represented this grievance to his Excellency Lord Loudoun.'[27] Captain John Ferrier's problems were more mundane. He had had to struggle to construct a durable bridge at Abrantes, and he wrote of the addition of 'boats, baulks, and planks, which I was obliged to make at each end of the bridge on account of the excessive rise of the river.' Given command of a Portuguese regiment at Olivença, Sir James Foulis found that he had only 'two battalions that scarcely make up 600 betwixt them, 700 militia and 50 valiant peasants mounted upon mares.'[28]

Loudoun's correspondence is full of complaint, and that was towards the end of a conflict, the Seven Years' War, in which the British had acquired considerable experience in overseas expeditions and in relations with allies. Many of the problems, especially the shortage of food and poor communications, reflect the difficulties of operating in a poor region, yet it is also clear that the army's ability to cope was limited, even for a rich power. Nevertheless, the army continued to operate effectively. The Bourbon (French and Spanish) retreat from Portugal in late 1762 can, in part, be attributed to the start of the winter rains and to the awareness that nothing could be gained in the peace negotiations, but the strength that the British brought to the Portuguese resistance had been important in preventing the Bourbon invasion from being successful.

Movement

More generally, military operations faced not only specific problems of campaigning in particular regions, but also the wider constraints posed by technological, economic and social conditions. Technological limitations were paramount. At sea, there was only a limited amount that could be done with wooden vessels subject to decay and dependent on windpower. Ship performance was depressed by the fouling of hulls: the luxuriant marine growth below water which was a particular problem in tropical waters, along with its kindred problem, attack on timbers by marine worms. The remedy of coppering was applied to the British fleet in the 1770s, in an impressive display of organisational capacity, though the French did not adopt it until 1785 (see pp. 151–2). Poor and unseasoned timber was another major problem, leading to ships being dismasted in storms. This made the supply of top-quality, generally Baltic, naval stores, and the denial of them to enemies, an important priority of diplomacy and strategy, the latter a case of employing force in order to deny capability to opponents.

The role of wind and tide ensured that strategically and politically desirable objectives, such as all-weather blockades or winter operations, were hazardous or impossible. It was with no mere form of words that the British Lords of the

Admiralty ordered Admiral Sir George Rooke to sail into the English Channel in April 1696 'with the first opportunity of wind and weather.' The weather affected both short-range and distant operations. The British fleet sent to take Gorée in West Africa from France in late 1758 lost three ships in a gale off the North African coast, while the expedition sent against Charleston in 1776 was delayed and dispersed by storms.

On land, there was no means of rapid communications for men or supplies. This had serious operational implications (as well as, by modern standards, tactical ones), leading, it has been claimed, to a crisis of strategy.[29] Larger armies posed serious problems, not least because there was no sustained increase in agricultural production, other than in a few parts of Europe, certainly in the first century of the period and in much of Europe until later in the nineteenth century. Furthermore, the difficulty of moving supplies helped focus sustained operations on fertile regions, such as Lombardy and the Low Countries. As these were also the best defended, this focus affected the tempo of operations. Logistical factors played a crucial role. Armies had to be supplied with fresh recruits, pay, food and munitions; and the stockpiling of magazines, for example by the French in 1740,[30] was therefore seen as a sign of warlike intentions. Pay was difficult to provide, but relatively easy to move, being of comparatively low bulk. Food and munitions, in contrast, were bulky and needed in vast quantities. Not only had the men to be fed, but also their horses and beasts of burden, such as oxen. Arms and artillery, shot and cannon balls were bulky and heavy, frequently getting stuck in the mud.

To move them was not easy. Europe lacked a system of roads, let alone good roads. Instead, there were a small number of well-developed routes, such as that from Paris to Lyon or the road following the Roman Via Emilia along the northern side of the Apennines from Bologna through Parma, which therefore played a key role in operations, as in 1734, and a mass of tracks of varying quality. Much depended on climate, soil and drainage. Impermeable soils, such as clay, quickly became quagmires after rain, but even routes on good soils could be hindered by poor drainage, heavy rains and snow melt. Bridges were infrequent and many rivers were only crossed by ferry. Wooden bridges and ferries, moreover, were easy to destroy in wartime. As a result, control over bridges could play an important role, both tactically and operationally, which helped explain the location and role of fortresses, not least in conjunction with logistical considerations. Thus, in 1726–7, and 1734, discussion about French moves across the Rhine into Germany focused on the crossing point at Rheinfels.

Furthermore, varying watercourses and flooding were problems in, for example, northern Italy. Flooding was a particular issue during the spring thaw, helping to make it a bad period for campaigning. The roads in eastern Europe were especially bad, and this contributed to the marked slowness of the Russian army on the march. The most important Russian road, that between St Petersburg and Moscow, was laid out by Peter the Great in the first two decades of the century. The road consisted of tree trunks, with piles driven into the marshes and low-lying soft spots. Covered with a layer of gravel, sand or dirt, such a roadbed was supposed to provide a firm and relatively smooth surface, but the rotting of the wooden base, erosion of the surface and gradual subsidence of long stretches into the soft, marshy soil, kept it in a permanent state of disrepair. Major Russian secondary roads lacked any roadbed and were simply a cleared expanse on which construction and cultivation were forbidden. Whatever the quality of the route, it generally could not stand up to the pressures created by a moving army, and ruts excavated by heavy wagon wheels aggravated the situation still more. This further lessened the mobility of artillery, especially the heavy siege cannon.

Rivers were not much better. Few rivers had been canalised, and many therefore suffered from variable water levels, weirs and tortuous meanders. Many ships had to be towed, never an easy operation in hilly country, and, due to the current, numerous rivers were only one-way routes. Nevertheless, they were cheaper than road transport. In 1703, the Swedish army used the Vistula to move its heavy baggage and artillery in Poland. As a separate issue, troops operating in waterlogged regions, such as the Danube valley in southern Hungary, were prone to serious disease. In the course of the century, canals were constructed in several countries, including France, Prussia and Russia, while Lombardy and the Low Countries possessed relatively good systems of water-borne transportation, but this was not a practical option in most of Europe. In addition, winter freeze, spring thaw and summer drought rendered most rivers seasonably undependable.

Most supplies had to come by wagon. Long supply trains followed those armies which received their supplies in whole or part from bases or magazines. It was calculated in 1744 that the siege train required by the Allied army in the Austrian Netherlands would 'amount to 10,000 horses and 2,000 wagons', at a cost of 50,000 pounds for six weeks.[31] The importance of wagon-borne supplies underlined the continued extent to which military operations depended on the availability of large numbers of horses. This ensured that fodder was an issue and that diseases affecting horses were a major problem. In 1748, Venetian contractors were unable to find the 5,000 mules required by the Austrian army in Italy.

Military service

If the quantity of supplies was a problem, so also was the quality. This was true in particular of food and arms. Poor-quality food affected performance to an extent that the available records leave unclear. Limited preparation, storage and transportation techniques, including the absence of refrigeration, led to much food being spoiled. Poor health was an obvious consequence, though the weakening of those soldiers, many already not especially strong, who were not listed as ill, might have been more serious in terms of such activities as marching. It was also important to be physically strong in order to carry and point a musket. Muskets were heavy and if not held level — musket droop — shot could roll out or fire could fall short. Serving cannon on both land and sea was also physically arduous.

More generally, poor health and, even more, epidemic disease had a major impact, particularly in tropical campaigns and in the crowded conditions of ships. Armies were mass transmitters and victims of illnesses, helping bridge different disease regions and thus exacerbating problems in areas with limited immunity to particular illnesses. The diseases that raged in the Russian camp at Narva in 1700 infected the Swedes when they seized it. Austrian troops brought illness to the Upper Palatinate in 1752 and from Hungary to Silesia in 1758, while the Russian troops operating in the Balkans during the 1768–74 war with Turkey spread typhus to Russia.

Improvements were made in military medical care, but, as the causes and means of transmission of several major diseases were not understood, there was a limit to what could be done. The value of ventilated buildings, warm clothing and bedding, and adequate food, was increasingly appreciated in the barracks that were built in the eighteenth century as civilian billeting became relatively less important (see p. 187), but these facilities were difficult to provide in the field or at sea. Troops on campaign often slept in the open air, which exposed them to the damp and lessened their resistance to disease.

No solution to the problem of ensuring that a standing army was kept up to strength without continual infusions of new recruits was found during the period, and this meant that soldiers were often inexperienced and judged unreliable. In turn, this encouraged training through drill in infantry tactics that stressed the general firepower contribution of a unit rather than individual shooting. This was further made necessary by the nature of the available weaponry. Although the production of standard firearms was greater than elsewhere, the weakness (by modern standards) of the Western industrial base and, in particular, its technological limitations, helped to produce serious problems with the reliability of weapons. Craft production meant a variation in specification, and thus that parts were not interchangeable. The difficulties created by short-ranged weapons, which had a low rate of fire and had to be resighted for each individual shot, were exacerbated by problems associated with poor sights, eccentric bullets, heavy musket droops, recoil, overheating and misfiring in wet weather. As guns, both muskets and cannon, were smooth-bore, and there was no rifling, or grooves in the barrel, the speed of the shot was not high, and its direction was uncertain.[32] Non-standardised manufacture (the calibre of individual Prussian muskets ranged between 18 and 20.4 millimetres, and their length varied by up to 8 centimetres), and wide clearances meant that the ball could roll out if the barrel was pointed towards the ground, while, at best, the weapon was difficult to aim or hold steady.[33] It was estimated in 1757 that many Russian muskets could not fire six times without danger of breaking.

Cannon were affected by muzzle explosions, defective caps and unexpected backfiring. There were improvements in the mid-eighteenth century, including the replacement of wooden by iron ramrods (by the Prussians in 1718) and better sights, but the situation remained one of poor individual accuracy for both artillery and infantry. As cannon could not fire indirectly, they could not be placed behind cover but had to be trained directly on their targets, although mortars and howitzers, with their high trajectories, were different. The absence of smokeless powder meant that firearms were badly affected by smoke, and, after the first shots, battlefield visibility was limited, which put a premium on the fire discipline required to delay shooting until a short range had been reached, an issue in the question of which side was to fire first at Fontenoy in 1745.

Poor visibility also affected command decisions on a number of occasions, and in 1758 Charles, 3rd Duke of Marlborough wrote to his wife, 'I must do my duty as a general, keep clear of the smoke and consequently out of shot to see what is going on in order to give proper orders.'[34] The absence of modern devices for night-fighting ensured that visibility was a particular problem at night, and this led to the end of many engagements as twilight gathered pace. Moonlight could therefore be important, a moonlit night in March 1702 being used for the successful invasion of Brunswick-Wolfenbüttel by the forces of Hanover and Celle.

These and related issues had major effects at both the tactical and the operational level. A stress on change and reform, whether on land or sea, can make the period appear one of continual improvement and progressive adaptation. In practice, as with other aspects of government activity, the degree to which military forces provided a good instance of the clash between aspiration and reality is striking, and it is important to note continued limitations in capability. Furthermore, policy was heavily reactive, with action taken in response to failure.

Although the limitations of administrative systems and ethos undoubtedly had serious military consequences, many problems could not be ascribed to them. One of the most serious was the extent to which land and sea operations generally took place for only half of every year, commonly from April until October. When the grass began growing, horses and other beasts of burden could be fed at the roadside. This was important as many armies, such as the Sardinian one operating in neighbouring Lombardy in March 1734, lacked magazines of dry forage. Even when such magazines existed they could become exhausted, as the British army in the Austrian Netherlands discovered in 1744. Their operations were affected as a consequence, while the attempt to obtain contributions from occupied French territory was unsuccessful because of the size of the demands. Forage was also to be a major problem for the French in Germany in the Seven Years' War.

After the spring thaw, roads and the land were usually reasonably firm until autumnal rains made routes impassable and filled siegeworks with water. Supplies were more plentiful in the late summer, when the harvest had been gathered in.

Winter, in contrast, posed particular problems, not least the lack of forage that ensured that in Portugal in the 1660s part of the cavalry had to be sent away from the combat zone. There were still examples of important campaigns then. In the winter of 1674–5, Austrian and Prussian troops fought a bitter campaign with the French in Alsace, while, in December 1677, the Swedish fortress of Stettin fell to Prussian forces after a long siege. However, such operations could be costly. Sleeping in the field with little or no cover, led to troop losses through death and desertion. Furthermore, winter operations were unpredictable, as in Italy during the War of the Polish Succession: in 1733–4, good weather allowed the French to cross the Alps and, in alliance with Charles Emmanuel III of Sardinia, to overrun most of Austrian-ruled Lombardy, the roads being passable for the French artillery. However, by February 1734, snow was hindering the French siege of Tortona, and both they and their Spanish allies lost many troops due to winter sicknesses. The following winter, the armies remained in their winter quarters, affected by a shortage of money and forage, and by sickness, while spring rain delayed the beginning of the 1735 campaign. In late 1742, heavy rain prevented the British forces in the Austrian Netherlands from advancing to attack the French, while a savage winter helped to decimate the French in Bohemia. Such problems helped discourage rulers from mounting winter offensives. In 1705, Louis XIV chose not to besiege Turin in the winter, postponing the siege until May 1706, by when, however, the defences were much stronger. In some parts of the Western world, for example Portugal and South Carolina, the heat of summer also discouraged campaigning.

Bad weather had a more serious effect at sea, where ships could lose their masts, rigging and cables in storms and be driven aground. Troops moved by sea in the winter generally suffered. Winter weather also made blockade far more difficult, a British diplomat reporting in 1756, 'the French say that His Majesty's [the British] fleet cannot hold the sea long [in the Mediterranean] as the winter is approaching, and that when Sir Edward Hawke retires into port, their communication will be again opened with Minorca without their risking an engagement.'[35]

The strains of winter operations, on land and sea, compounded the more general problems of military life. Alongside the support of comradeship, the excitement and the freedom of leaving home communities which could be as conformist and regimented as military life, there was not only the boredom of much of this life (not least drill), but also poor accommodation, the cold of drill as well as campaigning, as well as privations on campaign, including a lack of food, poor medical care and the dangers of the job.[36] The psychological and emotional experience of combat compounded the situation. These problems were faced even more seriously on the borderlands of the West, where troops often felt remote, isolated and neglected. Conditions were frequently poor. Thus, for the British army on the American frontier in the 1760s, diet and health were interlinked issues and chronic sickness and injury serious problems. The frontier force was worn down, displaying the effects of age and long service.[37] Similar remarks could be made about other frontiers, highlighting the need for caution in assessing the strength of what Geoffrey Parker has termed 'artillery fortresses',[38] and, more generally, in evaluating the relative capability of Western units.

4

WARFARE, 1660–88

Two major sieges were the most dramatic military episodes of this period. That of Vienna by the Turks in 1683 ended in total defeat and was a key moment in the balance of military power between the West and Islam. Four years later, in 1687, the fortress of Golconda (near Hyderabad in India) with its four-mile-long outer wall was besieged by the Mughal army under the Emperor Aurangzeb. The walls were bombarded with artillery while two mines were driven under them, although they exploded prematurely. The fortress finally fell by betrayal (a frequent result of sieges in India), Mughal forces entering through an opened gateway. This success led to the annexation of the sultanate, anchoring Mughal power in central India, a long-standing Mughal goal, although an achievement that was to be reversed by the mid-eighteenth century.

This siege by a large army, the supply of which was a formidable task, was a massive undertaking that puts in the shadow many Western enterprises of the period. So also did the major war in China in this period, the Sanfen Rebellion, or War of the Three Feudatories, which broke out in 1674. This was begun by powerful generals who were provincial governors, especially Wu Sangui, who controlled most of south-western China, but, having advanced, they were driven back to the south-west by 1677, although the rebellion did not end until 1681. It was the last major struggle within China until the Taipeng Rebellion of the mid-nineteenth century.

Reference to conflict in East and South Asia underlines the multiple military narratives that are possible for this period. This is particularly evident in China, where, alongside musketeers and cannon, archers and swordsmen played a role (unlike in the West), while, in the western Deccan in India from the 1660s, the mobile, lightly armoured Maratha cavalry proved a serious challenge to the Mughals, offering a parallel to the Polish and Hungarian use of light cavalry in Europe. This serves as a valuable reminder, not least when looking at the West, of the variety of effective forces that could exist in this period and the range of tasks that the military had to confront.

This has been underlined by recent work which has drawn attention not to the battles and sieges that tend to dominate scholarship, but to the 'small warfare' of the period, not least the role of raiding. At this level, there was an emphasis on surprise and march security, and on cover. Many actions took place around cemeteries, churchyards, villages and walled farms that one side had occupied for their defensive value. Most infantry actions in 'small warfare' were decided within minutes of the first volley. Far from being an inconsequential echo of battles, such conflict could serve operational goals: the systematic gathering of intelligence increased the possibilities of, and for, planning, while raids took on a meaning within the calculus of supplies and became a means of putting pressure on garrisons.

As an instance of the different ways in which established issues can be considered, this indeed offers a novel way to approach French war-making under Louis XIV (r. 1643–1715), the subject which tends to dominate scholarly attention for the period. Alongside the grand gestures of major operations and discussion of developments in tactical formations comes the need to consider the extent to which 'small warfare' successfully played an operational role, a point that can more generally be made not only on land but also at sea, with privateering in particular fulfilling operational goals. 'Small warfare' can be analysed, in the 1670s, in terms of the offensive-defensive character to French war-making on land, with the French trying to protect their northern frontier as well as to put pressure on the Spaniards, with whom they were at war from 1673 to 1678. It was impossible to seal off the frontier, but the French dispatched war parties, prepared defences, issued ordinances, cajoled local officials and organised militia, all with the intention of slowing and disrupting the Spanish imposition and collection of contributions and, thus, of lessening Spain's capacity to wage war and to retain its hold on the Spanish Netherlands (essentially modern Belgium and Luxembourg). In turn, as an aspect of anti-societal warfare, French reprisal raids and escalating demands for contributions were intended to inflict harm on the inhabitants of the Spanish Netherlands and to weaken their resolve. Spain sought to impose comparable pressure: the French defences worked well, but driving the Spanish garrisons from their fortresses was the only sure means to prevent Spanish raids.

Alongside the French pressure on the government and people of the Spanish Netherlands, there were more particular operational goals involved: blockades of fortresses were an important prelude to sieges, and they demonstrated the potency of French 'small war'. Blockades, which crucially continued during the period of winter quarters, in what, in effect, was a war of outposts, wore down the Spaniards, enervating their garrisons or ensuring that sieges were more rapid. The role of 'small war' alongside more major operations underlines the extent to which Western European warfare had similarities with that further east.[1]

'Small war' can be seen as an aspect of planned pressure, but was also a response to the desperate limitations of war-making in the period, not least the extent to which pay was often much in arrears, and troops therefore needed to raise their own funds. This can be seen in Portuguese and Spanish cross-border raiding in the long war of 1640–68, although it is also necessary to note the extent to which this drew on the traditions of the fighting that accompanied the reconquest of Iberia from the Moors and was a response to geographical constraints and possibilities.[2]

Alongside similarities with conflict in eastern Europe and, indeed elsewhere in the world, however, it is important to emphasise the variety that was western European military activity. This also means challenging the standard baton-exchange account with its focus, for the period covered in this book, on France and Prussia. That challenge can be enhanced by two additional shifts in the historiography. First, prefiguring the extent to which post-1815 accounts of the West as a whole need to give due weight to Latin America,[3] there is a different western Europe on offer. This includes a focus on Spain, no longer treated simply, as so often, as the sick man of Europe. It is worth considering how far Spanish resilience in the late seventeenth century looked towards the success of Spanish intervention in Italy in the first half of the eighteenth century, particularly the conquests of Sardinia, Sicily, and Naples and Sicily in 1717, 1718 and 1734–5 respectively, success that has received insufficient attention in the anglophone literature.[4] Subsequently, Spain was less successful in Italy in the War of the Austrian Succession, but its ability to mount an effort was again indicative of its continued military importance. The variety of western European developments can also be grasped by considering work on Italy,[5] while, within Germany, there has similarly been considerable interest in the 'Third Germany': states other than Austria and Prussia.[6]

Second, there is room for rethinking developments in the 'core' powers, especially France. Alongside work that stresses the relative sophistication, effectiveness, success and modernity of forces of the period,[7] comes studies that provide a different account, both of military developments and of the general socio-political situation. This work discerns a change during the century, but one of culture and mentalities, rather than due to military technological change, and it also draws attention to particular deficiencies in French war-making. For example, French operations in Italy in the 1690s were badly affected by the weaknesses of the supply system, and the same was true for operations against Spain in 1695. In addition, an emphasis on direct royal command of particular campaigns led to them receiving disproportionate resources, as dynastic prestige became the key strategic goal, at the cost of other strategic and operational considerations.[8] This was a long-standing problem with French war-making but also with that of other powers.

Far from focusing on change, this approach, and the discussion of limitations in French war-making, offers a degree of continuity with the situation earlier in the seventeenth century, not least in the important role of elite politics in constraining the choice of commanders.[9] While there were certainly changes from the situation in France in the 1630s, it is necessary to be cautious in discussing this in terms of a transformation, an argument that tends to be linked to a misunderstanding of the nature of 'absolutism', the term usually applied to the French political and governmental system. Irrespective of this concept of greater governmental power, which is, in practice, exaggerated,[10] the limited ability of the French crown to control operations, and, even more, the serious financial problems that affected policy, particularly in conflicts that lasted several years, in fact, emerged repeatedly. In the case of the navy, the situation under Cardinal Richelieu, Chief Minister from 1624 to 1642 – traditional methods of government through personal politics, rather than major institutional innovations[11] – changed to a certain extent from the late seventeenth century. However, it was still the case that the central agencies of the French navy had scant control of the, in effect, autonomous naval bases: the agencies could move levers, but, frequently, little happened at the bases.

The general misunderstanding of absolutism is related to the problematic use of the notion of modernity; and this is problematic not least as the variety of warfare in any one period acts as a qualification of any apparent clear-cut pattern of modernisation. As also at the scale of 'small war', the details do not, in fact, suggest much change from earlier conflict, whether or not seen in terms of modernisation. For example, the conditions under which Frederick William, Elector of Brandenburg-Prussia, the 'Great Elector' (r. 1640–88), led his forces in 1660 were not far different from those of the fifteenth and early sixteenth century.[12]

In assessing continuity, it is also appropriate to consider the goals of conflict. Alongside an emphasis on campaigning in order to defeat opposing forces, it is necessary to adopt a broader approach to war-making and victory and to underline the extent to which success had a symbolic value. From this perspective, decisiveness should be reconceptualised, away from an emphasis on total victory, understood in modern terms as the destruction of opposing armies and the capture of their territory, and towards, instead, a notion that may have had more meaning in terms of the possibilities and values of the period. This would note the extent to which wars ended in compromise peaces, and would present war as, in part, a struggle of will and for prestige. The ends frequently sought were, first, a retention of domestic and international backing that rested on the gaining of *gloire* and, second, persuading other rulers to accept a new configuration of relative *gloire*. This led to a concentration of forces on campaigns and sieges made important largely by the presence of the king as commander, such as Louis XIV's siege of Maastricht in 1673.

Differences in interpretation over the nature of French (and other Western) military systems in the late seventeenth century reflect in part the unresolved nature of the debate over an early modern military revolution. Aside from the problematic revolutionary character of the military changes of the sixteenth and early seventeenth century (the period of the supposed military revolution),[13] this thesis, however, makes only limited sense in terms of much revisionist work on sixteenth- and seventeenth-century Western history; and this is important for an assessment of issues such as goals, command and recruitment. In place of the traditional notion of new monarchies employing cannon and professional forces to monopolise power, this revisionist work emphasises the continued strength of the nobility and the value of a revival in the late seventeenth century of their consensus with the monarchs, the sociopolitical underpinning and expression of what is generally presented as absolutism. Indeed, in light of this scholarship, it is scarcely surprising that cavalry and noble officership both remained important in military culture and practice. Military roles and tasks played a major role in holding polities together, not least by linking landed elite and ruler through activity and shared purpose. At the same time, it is correct to complement the revisionist perspective with an awareness of the extent to which force, including new military forms, was employed by rulers in order to advance their power and enforce their authority. Thus, in 1660, Louis XIV had Marseille occupied by royal troops and a section of the city wall demolished, a practical and symbolic demonstration of royal control and the limitations of urban autonomy.

It would be mistaken to imply by any periodisation that all armies and navies waging war were developing in a uniform and consistent fashion, although, in the late seventeenth century, the armed forces of the major Western states grew appreciably, and there were also important tactical developments. Initially, however, there was a slackening of tension in the 1660s, and a post-war reduction of much military strength, not least in England and France. The end of the Interregnum regime with the restoration of the Stuarts, in the person of Charles II in 1660, was followed by a marked reduction in the size of the English army as a result of domestic political exigencies, particularly the linked distrust of the possible consequences for liberty of a large army and a financial settlement that left the crown with limited financial resources. Furthermore, the lengthy and bitter struggle between France and Spain that had been continuous since 1628, with formal hostilities continually since 1635, ended with the Peace of the Pyrenees of 1659, a settlement in which France made important territorial gains: Artois and Roussillon. Peace between Spain and France's ally England followed in 1660, while the Treaty of Oliva (1660) ended conflict between Sweden, Poland and Brandenburg (later known as Prussia); that of Copenhagen (1660), the war between Denmark and Sweden; and the Treaty of Kardis (1661) marked Sweden's continued success in preventing Russia from gaining a Baltic coastline.

War over control of Ukraine continued between Russia and Poland, begun in 1654 and revived in 1658 after a settlement in 1656, until 1667, when a partition was accepted, and it was only in 1668 that Spain conceded independence to Portugal, ending the struggle that had begun with the Portuguese rising in 1640 and that became more large-scale from 1660 as both sides deployed larger forces, while the Portuguese benefited from French and, even more, English troops.[14] However, the essential agenda of conflict earlier in the century, much of which can be summarised by reference to the Thirty Years'

War (1618–48) and the civil conflicts collectively known as the mid-seventeenth-century crisis, had been brought to a close. As a result, across much of Europe, there was a reknitting of the relationship between rulers and significant sections of the social elite which earlier had been sundered, both by hostility stemming from the sixteenth-century Protestant Reformation and also from tensions within the composite monarchies, particularly 'Austria', 'Britain' and 'Spain', assembled by marriage and conquest. This reconciliation brought new political stability to many states, especially Austria and France, which had important military consequences. Nobles sought to win prestige and display influence with the monarch by raising and commanding troops, while stability eased the flow of money and credit, which was important as this was a period of economic depression. Whereas a significant proportion of the military entrepreneurs who raised and commanded units prior to 1660 were non-noble, the period after 1660 saw their widespread replacement by the nobility.

It is difficult to be precise about the number of troops raised across the Continent. Although governments sought to obtain precise figures, and paid wages and allowances accordingly, it was difficult (both then and now) to have much confidence in the figures available. Officers who wished to avoid problems, to emphasise their importance, and to embezzle pay, overlooked death, disease and desertion in order to present their units as being up to strength. Furthermore, there are problems in assessing the figures available, for it is rare that like can be compared with like. The numbers of those under arms (as well as the aggregate tonnage of warships) were not the same as the numbers of effectives, and the latter could and can be assessed differently, with major contrasts between sources. George III of Britain warned in 1800,

> as I have [observed] on many former occasions . . . I trust attention will be had to the real state of our forces, not to fallacious states on paper which always make a greater appearance than can be depended upon when the troops are collected, and consequently that we may not trust on having a larger force than can be effected.[15]

In addition, the absence of precise population figures makes it impossible to arrive at a precise percentage of the population under arms.

Despite these problems, there is no doubt that the numbers of those under arms in the West rose, a process that peaked in the last years of the Napoleonic wars. Gustavus Adolphus of Sweden had only about 30,000 men in his army in 1631, but, by the 1700s, one of Louis XIV of France's field armies could be three times as large, although the average size of individual field armies did not grow at the same rate. In most Western countries, this increase began in the late seventeenth century. Earlier increases had been important, but, in the late seventeenth century, increases in the size of armed forces were sustained. There were post-war demobilisations, but most rulers felt the need and the ability to retain larger standing (permanent) military and naval forces than they had prior to 1660. The late seventeenth century was the period in which many states created standing armies.

Nevertheless, it would be mistaken to suggest that there was a consistent relationship in the size of armed forces between different states, or, indeed, that all states had a rise in numbers. Many states, such as Bavaria, reduced their forces after the War of the Spanish Succession ended in 1714, and did not exceed the totals raised in 1689–1714 until the 1790s. For example, the Hanoverian army, established in 1665, was increased with the Dutch War (1672–8), and thereafter averaged 10,000–16,000 men, rising to 22,000 men during the War of the Spanish Succession, but the establishment was reduced in 1715 to 14,500–15,00, and did not rise again to about 19,000 until after 1727.[16] Furthermore, although the strengths of the Austrian, Prussian and Russian armies increased after 1763, in most other states, including France and the majority of German and Italian states, they decreased after the Seven Years' War (1756–63), and did not rise again until the outbreak of the French Revolutionary War in 1792. States that were neutral in the Seven Years' War, such as the United Provinces and the Italian states, did not have an increase then and, instead, their forces decreased after the War of the Austrian Succession ended in 1748.

Increases in forces were seen on both land and sea in the late seventeenth century. They characterised both the total numbers in armed forces and the numbers of those put in the field. Furthermore, any increase in forces, whether on land or on sea, had a cumulative effect: potential opponents felt it necessary to match increases, because the size of an army or navy was a source of prestige, and was also believed to be strategically and operationally effective.

Specific military developments also encouraged an emphasis on numbers in this period. Line-ahead naval tactics led to heavier broadsides and, thus, to a need for larger warships that encouraged a 'naval race' between England, France and the United Provinces, as well as competition between Denmark and Sweden in the Baltic. Larger ships meant bigger scantlings. The effectiveness of numbers, especially of heavily gunned warships, increased, while it was believed that the larger of two opposing fleets would be able to attack the rear of its opponents. Interlinked and mutually sustaining changes led to an increase in naval force. Firepower was enhanced by the use of cast-iron guns, which were cheaper and sufficiently dependable to replace the much more expensive, but also lighter, bronze guns. Heavier shot was fired, while English broadside firepower increased with the development of improved tackles which used the gun's recoil to speed reloading inboard. The essential resilience of wooden ships ensured that they were difficult to sink by gunfire, but cannon firing at short range could devastate rigging and masts and effectively incapacitate the ships.

The growth in Western naval power was not simply a matter of developments afloat. New naval bases were created and existing ones enhanced. Whereas, in 1663, half of France's naval expenditure had been on the galley fleet,[17] reflecting the importance of competing with Spain in the Mediterranean, a new geography of naval power emphasised the Atlantic. The French developed Brest as their major naval base, and the English in the late seventeenth century increasingly shifted from an emphasis on Chatham, on the North Sea, to Portsmouth and Plymouth, that is, from opposing the Dutch, as in the three Anglo-Dutch wars of 1652–74, to confronting France, the goal from 1689. The French fleet greatly increased in size in the late 1660s and early 1670s, so that by 1675 it was the largest in the world.

France

The major increase in the size of the French army from the 1670s was neither cause nor consequence of tactical shifts. Instead, it is appropriate to emphasise the stronger domestic position of the French monarchy under Louis XIV and the extent to which the army lent weight to Louis's assertive, and frequently aggressive, foreign policy. The French army increased in size because Louis believed this necessary to give weight to his policy and could obtain the soldiers, not because technological or tactical innovations encouraged any such change, while the greater size posed formidable logistical problems. A growing awareness of the limitations of the French army during the administrations of Cardinals Richelieu and Mazarin (1624–61) throws Louis's achievements into greater prominence. Wartime peaks and peacetime figures both rose. These increases were a response to Louis's ambitions and to his growing international isolation. In contrast, in 1635–48, the French had been greatly helped militarily by alliances with the Dutch and the Swedes against Spain and Austria respectively.

An estimate of real wartime size (as opposed to paper strength) puts the number of troops under arms as rising to a maximum of 125,000 in the conflict years of 1635–48, 253,000 in the Dutch War of 1672–8, 340,000 in the Nine Years' War of 1688–97, and 255,000 in the War of the Spanish Succession of 1701–14. In contrast, the figure for 1794 was 750,000 men, but the numbers raised under Louis XIV were far greater than those under his predecessors and represented the maximum that his resource base could support. The number declined in the 1700s because, first, the French were having to fight increasingly at home, and could not therefore offset costs through 'contributions' from occupied territory; second, the decade was difficult economically, with particularly poor harvests; and, third, as royal and aristocratic finances had been exhausted in the 1690s.[18] The peacetime strength in the French army in 1679–88 was about 130,000–150,000, although its potential was enhanced by the major investment in new fortifications.

During Louis XIV's reign the administration of both army and navy was improved, although, as was common with early modern 'reform' initiatives, it proved difficult to sustain change. In reaction to the chaos of the *Frondes*, particularly aristocratic opposition in the 1640s and 1650s, Louis and his ministers sought, with a measure of success, to gain control over the army, and to re-establish the principle and practice of royal direction. They had greater control over military commanders than Louis XIII (r. 1610–43) had done, although the limitations of governmental direction was to be seen after Louis XIV died in 1715, not least during the unpopular War of the Quadruple Alliance in 1719–20, while there were serious disputes within the military leadership during the next war, that of the Polish Succession (1733–5). Nevertheless, under Louis XIV, the payment of troops was regulated; drill, training and equipment were largely standardised; and in 1670 distinctive uniforms chosen by Louis were introduced, asserting royal control of the army and making desertion more difficult. The name of the Inspector-General of the Infantry (from 1667) and Commander of the Régiment du Roi, Martinet, entered the language as the description of a strict disciplinarian. The command structure was revised and a table of rank established in 1672, and, if social hierarchy continued to conflict with military seniority, the context was one of a relationship between crown and elite in which the property rights of officers to their position did not extend to any challenge to royal sovereignty.[19]

Furthermore, the French supply system, of men, money, munitions and provisions, was considerably improved, as was the system of *étapes* (depots along marching routes, where troops could obtain supplies) that supported it. Similarly, a network of magazines, from which campaigns could be supplied, was established near France's frontiers and was used with considerable success in launching the War of Devolution in the Spanish Netherlands (Belgium) in 1667 and the Dutch War in 1672. Thanks to the magazines, French troops were able to seize the initiative by beginning their campaigns early. Their conditions were also improved, not least with more reliable pay, clothing, food and medical services. These achievements owed much to two successive secretaries of war, Michel Le Tellier and his son, Louvois, who, between them, held office from 1643 to 1691, and who were supported by a well-organised war office. The navy was developed by Jean-Baptiste Colbert, Secretary for the Navy 1669–83. He organised a system for the conscription of sailors, purchased warships and recruited skilled Dutch shipbuilders, founded training schools, developed dockyards and administrative systems and saw a major expansion in the size of the navy.

Louis's army was spectacularly successful on a number of occasions, certainly far more so than the French had been against Spain and her allies in 1625–59. The campaigns in the Spanish Netherlands and Spanish-ruled Franche-Comté (the region round Besançon) during the War of Devolution (1667–8), in pursuit of Louis's claim on the inheritance of Philip IV of Spain (r. 1621–65), were considerable successes, as were the initial stages of the invasion of the United Provinces in 1672 during the Dutch War (1672–8), and the subsequent conquest of Franche-Comté in 1674. The closing campaign of the Dutch War was conspicuously successful, while the speed of French success was demonstrated in 1667, when Lille rapidly fell and the Spanish army was defeated at Bruges. It is possible to point in France's military history under Louis XIV to many deficiencies, to strategic overreach and operational mistakes and to note continued administrative limitations, not least, by the 1700s, a financial system that was far weaker than that of Britain; but the army, navy and military administration were more effective than in the past. The army was also used successfully to put down revolts, especially, but not only, in Gascony (1672) and Brittany (1675), and, in this respect, the situation was more favourable for the government than it had been under Louis XIII.

Moreover, France's relative military situation improved as a consequence of heavy emphasis on, and investment in, fixed positions: naval bases – Brest, Dunkirk, Lorient, Rochefort, Toulon and Quebec – and fortresses. It was only under Louis XIV that a major programme of new fortifications was pursued: under Louis XIII, there had been major works, for example at Pinerolo, but nothing that compared with the systematic attempt to defend vulnerable frontier regions with new fortifications that his son pursued. Appointed Commissioner General of Fortifications in 1678, Sébastien Le Prestre de Vauban supervised the construction of thirty-three new fortresses, including Arras, Ath, Blaye, Lille, Mont-Dauphin, Mont-Louis, and New Breisach, and the renovation of many more, both in France and in newly annexed territories, including Belfort, Besançon, Landau, Montmédy, Strasbourg and Tournai. In essence, Vauban's skilful use of the bastion and of enfilading fire represented a continuation of already familiar techniques, particularly layering in depth, and he placed the main burden of the defence on the artillery; but it was the crucial ability of the French state to fund such a massive programme that was novel.

Territorial expansion was directly linked with the construction or enhancement of fortresses, which were designed to consolidate acquisitions, and yet also to facilitate opportunities for fresh gains by increasing France's presence in contested areas, and safeguarding bases for operations, not least where stores could be accumulated. Thus, fortifications at Strasbourg and New Breisach helped both provide a defensive frontier for Alsace, and also control bridging points over the Rhine that permitted the projection of power into south-west Germany. Their ability to do both sustained uncertainty about Louis's goals.

Proficiency in siegecraft also increased markedly. Vauban directed forty-two sieges, including that of Lille in 1667, the key victory in the War of Devolution with Spain, in which French forces also captured other fortresses in the Spanish Netherlands, including Ath, Bergues, Charleroi, Courtrai, Douai, Furnes, Luxembourg, Oudenaarde and Tournai. The contrast with much of the French campaigning in 1635–59 was instructive. Lille was retained in the peace and a new citadel completed in 1670. In the next conflict, the Dutch War, Vauban demonstrated his skilled siegecraft, at the expense of Dutch-held Maastricht in 1673, showing how trenches could march forward by successive parallels and zigzag approaches, designed to minimise exposure to artillery and sorties and to advance the positions from which besieged fortifications could be bombarded and attacked.

The Dutch War

Tactical proficiency, however, was not enough in a conflict that did not develop as Louis XIV had anticipated. Angry with the Dutch role in the anti-French Triple Alliance (with England and Sweden) of 1668, which was designed to intimidate France into limiting its gains from Spain in the War of Devolution, Louis was influenced by the prospect of easy alliances and quick victories held out by Turenne, one of his leading generals. However, Louis had little sense of how a new war with the Dutch, once begun, would develop diplomatically or militarily. Louis hoped that Spain would come to the aid of the Dutch, and that he could therefore resume the conquest of the Spanish Netherlands. Having gained the alliance of Charles II of England (who controlled an important fleet), and most of the leading German princes, Louis attacked in 1672.[20] His army outflanked the Dutch by advancing through the bishopric of Liège and the Electorate of Cologne, crossing the Rhine and advancing into the United Provinces from the east. French forces seized Utrecht, but their advance on the province of Holland was stopped by flood-water when the Dutch opened the dykes.

In 1673, the war broadened out as the Holy Roman Emperor Leopold I, the ruler of Austria, organised German resistance. The Austrian general, the talented Count Montecuccoli, who had defeated the Turks in 1664 at St Gotthard, outmanoeuvred Turenne and forced him to retreat from east of the Rhine, and French-garrisoned Bonn, the capital of the allied Elector of Cologne, was captured. Under pressure, Louis abandoned his position in the United Provinces in 1674 in order to protect the French frontier, while Charles II and the Prince-Bishop of Münster, who had attacked the Dutch from the east, came to terms with them that year. The entry of Spain into the war in 1673, however, allowed French forces to advance in the Spanish Netherlands, where, in 1674, they were victorious at Seneffe, to invade Catalonia, and to conquer the Franche-Comté – its major positions, Dôle and Besançon, falling in 1674. Also that year, Turenne shored up the French position in and near Alsace, winning victories over the Austrians at Sinsheim and Enzheim, and capturing Strasbourg, before staging a surprise winter campaign in which he was victorious at Mulhouse and Turckheim, clearing Alsace.

Campaigning against Montecuccoli in 1675, Turenne was killed by a cannonball in battle at Sasbach, a reminder of the dangers to which generals were exposed. Montecuccoli advanced deep into Alsace, only to be driven back across the Rhine. In 1676–8, the French Marquis of Créqui and his Austrian rival, Duke Charles of Lorraine, campaigned in the Rhineland, with mixed results. Austrian forces successfully besieged Philippsburg and invaded Alsace, but were defeated at Kockersberg (1677) and Rheinfeld (1678). Like the French victories in the Low Countries at Mont Cassel (1677) and St Denis (1678), these are now forgotten encounters, but they reflect the extent to which the conflict of the period was high tempo and involved the uncertainties of battle, as well as the more certain processes of siege, which the French brought to a fine art, their captures in the Spanish Netherlands including Dinant, Huy and Limburg in 1675, Condé, Bouchain and Aire in 1676, Valenciennes and Cambrai in 1667, and Ghent and Ypres in 1678. In contrast, William of Orange was unsuccessful at Maastricht (1676) and Charleroi (1677). The number of these sieges reflected the heavily fortified nature of the region. Battles were frequently related to sieges, as attempts at the relief of besieged positions were resisted.

The war was far from restricted to the oft-contested lands between the North Sea and the Alps. There was also an important naval dimension, the range of which indicated the number of combatants. Off Sicily, French ships sent in 1674 to help the revolt of the city of Messina against Spanish rule, another instance of a long-established policy of exploiting discontent in the Spanish empire, clashed with the Spaniards and Dutch, with both inconclusive battles (Lipari Islands, 1676) and French victories (Messina, 1675, Augusta, 1676, and Palermo, 1676). The French also sent an expeditionary force to Sicily that campaigned from 1675 to 1678, although without conquering the island. Instead, as in the conflict of 1635–59, Spain retained its position in Italy and proved able to thwart French hopes of major gains. The difficulties of army–navy coordination and cooperation were more generally seen in this period, for example in the failure of the Spanish attacks on Portugal in 1666.

In the North Sea, the English attacked the Dutch jointly with France, in an attempt to assist the French invasion of the United Provinces. The Dutch, however, successfully fought off their more numerous opponents in naval battles such as the Texel (1673): poor Anglo-French cooperation was a factor. Further afield, the Dutch attacked English and French shipping in the Caribbean and recaptured New York, which they had lost to the English in 1664. The city was returned in 1674 when Charles II abandoned his alliance with France. A significant French fleet was sent to the Indian Ocean, but attacks on the Dutch in Sri Lanka and southern India failed, although a base that was to be the basis of French power in India was established at Pondicherry in 1674. However, in India, an alliance of Golconda and the Dutch forced the French at Masulipatam to surrender that year, and, in the West Indies, the French expedition against Curaçao fatally ran aground in 1678.

Naval conflict spread to the Baltic where Denmark (on the Austrian side) and Sweden (on the French) fought from 1675 to 1679. This was the last war in European waters in which armed merchantmen were much used. The more powerful Danish fleet was victorious at Moen (1677) and Koge Bay (1677), but a Danish invasion of Sweden, that attempted to reverse the Swedish conquest of Scania (southern Sweden) earlier in the century, was defeated at Lund in 1676. The previous year, a Swedish invasion of Austria's ally

Brandenburg-Prussia was defeated at Fehrbellin, a relatively minor battle that, nevertheless, came to play a major role in the Prussian historical imagination.[21]

Repeated victories in the Dutch War helped establish an image of French military proficiency, not least in their own mind. The *Galerie des Glaces* at Versailles, the largest long gallery yet built, had ceiling paintings commemorating in particular triumphs during the recent war.[22] In part, however, French success owed much to the unstable nature of the opposing coalition. French diplomacy played on the divisions among her opponents, ensuring that, in the Peace of Nijmegen of 1678, there was no single treaty ending the war, but, instead, a number of abandon-the-allies agreements that enabled the French to gain reasonable terms, not least a more defensible frontier. The impact of military strength was shown when Frederick William of Brandenburg-Prussia was obliged in 1679 by French military moves to return his gains from Sweden.

The Dutch War also pushed forward military developments in other states, encouraging investment in standing (permanent) forces, and thus the accumulation of experience and the embedding of improvements. This was true of the Dutch and of several German principalities, including Brandenburg-Prussia, which, from 1701, was to be a kingdom and is then best known as Prussia. Dutch forces rose from 30,000 after 1648 to about 100,000 in 1673, and, after a peacetime fall reflecting political divisions, to under 40,000 in 1685 to over 100,000 in 1702–12. Such forces were far more under the control of the rulers than had been the case during the Thirty Years' War (1618–48). Military contractors were brought under control, in part by integrating units that were hired into the regular command and administrative structures.

The extent to which the French continued to use force after the Dutch War was over accentuated this process and reflected the extent to which force was a habitual response in disputes. Louis XIV advanced a series of claims in a unilateral fashion and occupied territory accordingly in the early 1680s in both the Spanish Netherlands and the Holy Roman Empire, where Strasbourg was occupied in 1681. Intimidation seemed an end as much as a means and also helped provide supplies for the army. In 1683, for example, troops were sent into Spanish Flanders, ordered to live off the land, and Spanish responses led to an escalation of intimidation.[23] The navy was also used to blockade Cadiz in 1682 and 1684, and in a devastating bombardment of Spain's ally Genoa in 1684. Although Louis was in part motivated by concerns about French vulnerability, the overall character of his policies was aggressive.

Eastern Europe

Conflicts in western Europe led to governmental and political developments and greatly affected post-war diplomacy, but they can take attention away from the important, but more varied, conflicts in eastern Europe in this period. To a greater extent than further west, these conflicts saw not only clashes between different military systems but also the interaction of rebellion with foreign war. This was particularly the case in Poland, where a civil war in 1665–6 lessened the ability to pursue the conflict with Russia, leading, in 1667, to the Truce of Andrusovo, in which Ukraine was partitioned between the two powers. Similarly, the Austrian attempt to challenge the Turkish domination of Transylvania led to war in 1663. At the battle of St Gotthard (1664), an Austrian force under Montecuccoli stopped the Turks advancing across the river Raab, indicating that they were no longer dependent on fortresses in order to thwart Turkish advances. The war, nevertheless, was swiftly ended by the Peace of Vasvar, reflecting Leopold I's lack of confidence about any forward policy. The Austrian army was clearly capable of defensive operations, but there was doubt about the ability of both Austria and its army to sustain a major war with the Turks, doubts which were not to be satisfied until two decades later.

In 1671, the Turks attacked Poland, advancing their own claim to Ukraine, the rebellion of which against Polish rule in 1647–8 had helped provoke the confrontation between the major eastern European powers. Initially very successful, the Turks, under Sultan Mohammed IV, captured the major fortress of Kamienic Podolski on the River Dniester in 1672, turning its cathedral into a mosque where the Sultan prayed. The Peace of Buczacz recognised Turkish suzerainty over western Ukraine, but John Sobieski, King of Poland from 1674, renewed the war. Turkish defeats ensured that they were unable to sustain the position gained in the peace of Buczacz, but, by the Peace of Zuravno (1676), the Turks still emerged with the gain of the province of Podolia. Like the final end of Venetian resistance on Crete in 1669, this indicated the vitality of the Turkish military system.

Turkish concern about Russian control of eastern Ukraine led to its invasion in 1677. The major fortress of Chigirin on the River Dnieper was besieged but resisted long enough to permit the arrival of a relieving force. The following year, however, the Turks advanced again, and, this time, the Russians were forced to evacuate Chigirin, blowing up the fortress. Traditional Russian musketeer units, the *streltsy*, fared badly in these campaigns, and the artillery, though numerous, was ill equipped and mismanaged. But the Russians fought hard and limited the Turkish advance, so that, in the Treaty of Bakhchisarai (1681), they retained their control over the eastern Ukraine, while recognising Turkish rights to western Ukraine.

The Turks were already planning to concentrate their resources against Austria, exploiting, as so many states did, dissidence, a process eased by the extent to which states, both monarchies and republics, were really composite polities. A revolt in those parts of western and northern Hungary ruled by the Habsburgs encouraged the Grand Vizier, Kara Mustafa, to begin aiding the rebels, and to recognise their leader, Count Imre Thököly, as Prince of Middle Hungary; he, in turn, agreed to become a vassal of the Sultan, and thus to extend the system of Turkish buffer states which already encompassed Moldavia,

Transylvania, Wallachia and the khanate of the Crimean Tatars. In 1682, Turkish forces operating in support of Thököly captured Kosice.

At the head of about 100,000 troops, Kara Mustafa appeared before Raab (Györ), the key to the Austrian military frontier, on 2 July 1683, but, instead of besieging it and other fortresses near the frontier, a measure which would have greatly delayed him, he decided, without the approval of the Sultan, to march directly on Vienna, a step encouraged by French agents. Leopold I and his court fled, and the defence of the city was left to Count Ernst Rüdiger von Starhemberg, who had the suburbs burnt to deny the Turkish troops cover. The Turkish army surrounded the city on 16 July and began building siegeworks, and they also launched a series of increasingly successful assaults. The garrison suffered heavy casualties, as well as losses from dysentery. The Turks suffered similarly, but, during August, the city's outer defences steadily succumbed, although, lacking heavy-calibre cannon, the Turks were outgunned and forced to rely on undermining the defences. There was bitter fighting in the breaches and, on 4 September, Starhemberg fired distress rockets to urge the relief army to action.

This international force, commanded by John Sobieski, had advanced through the Vienna woods. Kara Mustafa was aware of this deployment, but made scant preparations for an attack. On 12 September, the relieving army descended from the hills and heavily defeated the Turks. As was common in such cases, Kara Mustafa was to be punished for his failure by strangulation, while Christian Europe celebrated the relief of its most prominent bulwark. Divisions among the Turkish forces (particularly the Khan of Crimea's distrust of Kara Mustafa) contributed to their defeat, as did serious command faults, and the Turkish right flank abandoned the battle.[24] However, this was not only a battle that the Turks lost. In addition, their opponents' success owed much to an improvement in tactical proficiency compared to earlier conflicts with the Turks, while they were also numerically superior.

The battle was followed by a major shift in power in Europe, and one that was to have lasting geopolitical consequences. Buda, a strong fortress which was the key to Hungary, successfully resisted a four-month siege in 1684, in part due to the problems the Austrians faced from disease and logistics, but also reflecting their limited capability in siegecraft.[25] However, Nové Zamsky fell in 1685, and, when Buda was besieged anew in 1686, a shell landed on the main powder magazine, blowing open a breach in the walls. Repeated assaults then led to the fall of the city. It was never regained by the Turks. The following year, Charles of Lorraine defeated the Grand Vizier, Suleiman Pasha, at Berg Harsan (Harkàny), moving on to capture Esseg. In 1688, the Austrians pressed on to capture Szegedin and Belgrade. As advance Austrian units moved further afield, some Christians in the Balkans talked of liberation, and the collapse of the Turkish position there appeared imminent. Meanwhile, in Greece, an anti-Turkish revolt in the Morea (the Pelopennese) had been exploited by Venetian amphibious forces in 1685–6, and, in 1687, the Venetians moved north to capture Athens, a mortar shell blowing up the Turkish magazine in the Parthenon. However, Russian failures against the Crimea in 1687 and 1689 indicated the grave logistical challenges of steppe warfare. The war continued until the Treaty of Karlowitz of 1699 (see pp. 65–6), contributing greatly to the prominent role of conflict in the late seventeenth century.

Britain

The decisiveness of military operations was further demonstrated at the opposite end of Europe where, in 1688, responding to the opportunity created by the unpopularity of the Catholic and autocratic James II of England (r. 1685–8), his nephew and son-in-law, William III of Orange, the leading Dutch politician and an experienced general, invaded England. This was the most effective amphibious operation of the period, but it also saw the key role of political factors. Concerns within England about James's policies helped undermine the latter and limit the military response to William both at sea and on land, compounding James's failure of nerve; and William was able to seize London without fighting. There was to be serious subsequent conflict in Scotland and Ireland (1689–91) as the new order was successfully established across the British Isles, but the 1688 campaign was not only an important qualification to the oft-repeated theme of the indecisive character of the warfare of the period, but also one of the most influential campaigns of the period.

Alongside the hard-fought defeat of the Turks, this campaign helped create the geopolitical configuration of Europe for the remainder of the period of this book, limiting the options for France, however successful its forces might be in the Low Countries, Italy and Germany. William III, who ruled Britain from 1689 to 1702, was very opposed to Louis XIV, and French support for the exiled Stuarts – the Jacobite movement (named after the Latin for James), helped add ideological conflict to political rivalry. France therefore faced two powers, Britain and Austria, which were only ready to cooperate with her, as Britain did in 1716–31 and Austria in 1756–89, if she agreed not to expand territorially in Europe, and, conversely, which allied against France when they judged her a threat, joining in war with France in 1689–97, 1702–13, 1743–8, 1793–7, 1798–1801, 1805, 1808–9 and 1813–15. This geopolitical configuration was to be cemented by the success of Russia discussed in the next chapter. As a great power, Russia was to oppose France's protégés in eastern Europe: Sweden, Poland and Turkey, becoming a great power as a result of its victories over them.

The wider world

Thus, multipolarity was secured in Europe. In 1660–88, this also still appeared possible in East Asia, as Manchu China had not yet achieved the dominance of its neighbours it was to secure by 1760. Nevertheless, Manchu success in the War of the Three Feudatories, alongside Aurangzeb's ability to establish Mughal power in the Deccan at the expense of Golconda, indicated the military-political proficiency of non-Western powers; as was also shown, albeit in the event far more shakily, by the ability of the Turks to press their neighbours, especially in 1663–4 and 1677–81, and to project their strength to the walls of Vienna. That the situation was totally different at sea underlines the difficulties of making any overall judgement of capability.

Outside Europe, Western forces had very varied fortunes. They were least successful in the Far East. In 1661, Zheng Chenggong, a Ming loyalist known to the Europeans as Coxinga, turned his attention to Formosa (Taiwan), invading with a force of 300 junks and 25,000 men. After a nine-month siege, he took the Dutch base of Fort Zeelandia, helped by the defection of some of the garrison, who then explained how it could best be attacked. Coxinga's success owed much to old- and new-style weaponry. He used his shield-bearers as an aggressive assault force, but several infantry attacks were beaten off. Coxinga also had twenty-eight European-style cannon, some directed by Dutch renegades, and the fort surrendered after the walls of its Utrecht redoubt collapsed under heavy fire. Dutch relief attempts from Java failed due to poor leadership, bad weather and insufficient troops; and squadrons sent in 1662–4 to re-establish the Dutch position were all unsuccessful.[26] Further north, in 1685, Chinese forces successfully besieged the Russian fortress of Albazin in the Amur Valley. The Russians rebuilt it, only to lose it to a second siege in 1686. In 1689, a Chinese advance as far as Nerchinsk led the Russians to acknowledge Chinese control of the Amur Valley. Furthermore, in 1688, the French garrisons in Siam (Thailand) were expelled after the pro-French ruler was overthrown in a coup.

In contrast, there were successful uses of force. In order to protect their Mediterranean trade, the French heavily bombarded Algiers in 1682, 1683 and 1688, and Tripoli in 1685, and also intimidated Tunis by the threat of bombardment. Further afield, in 1675–6 in New England, British settlers overcame native opposition in King Philip's War. About 7,900 of the 11,600 Native Americans living in southern New England on the eve of the conflict died, through battle, disease or exposure, or were removed by being sold as slaves or becoming permanent refugees.[27] The varying fortunes of war therefore were very much moulding the extent of the Western world.

5

WARFARE, 1689–1721

The focus in Western conflict in this period is in large part on warfare between largely similar forces, 'symmetrical warfare', although the large-scale wars between the Turks and their Austrian and Russian neighbours were very different in type. In Asia, the key conflicts led to the defeat of the Zunghars of Xinjiang by Chinese forces, and the Chinese conquest of Tibet, as well as the faltering of Mughal India, largely at the hands of the Marathas. In each case, there were major military contrasts between the combatants, but political issues were also significant in helping explain success and failure. In Europe, the most significant struggles were the defeat of the Turks by the Austrians, and, to a lesser extent, the triumph of Peter the Great of Russia over his Swedish rival, Charles XII, in the middle stage of the Great Northern War (1700–21). Conflicts between Louis XIV of France and his neighbours tend to attract more attention and are indeed of consequence as Louis's hopes of dominating western Europe were dashed, but they should not be seen as defining the nature of Western conflict in this period. The period 1689–1721 had the heaviest fighting between 1618–48 and 1792–1815, and, given the demographic growth from the 1740s, fighting then often had proportionately greater impact than in 1792–1815. Thus, the years 1689–1721 are central to the challenge to the cliché of *ancien régime* limited warfare.

The Balkans

At the outset of the period, the Austrian advance into the Balkans led to bold hopes about Turkish collapse, but in 1690 the Turks displayed the resilience that frequently characterised their efforts. The able new grand vizier, Fazil Mustafa, who had restored order to the army, mounted a counter-offensive. Benefiting from the withdrawal of most Austrian troops to face France, and the resulting changes in the tempo of operations, he retook Nish and, after only a six-day siege, Belgrade: timorous command, treachery by a French engineer and explosions in the powder magazine have all been blamed for its fall. In the following year, however, Fazil Mustafa's hopes of recapturing Hungary were dashed by Ludwig Wilhelm of Baden at Zálákemén, a long, hard-fought battle which left a third of the imperial army killed or wounded, but the Turks routed and Fazil Mustafa dead.

Conflict in the Balkans over the next few years was indecisive and difficult due to the unwillingness of either power to offer acceptable terms and to rival imperial commitments against Louis XIV: the Austrians fought alongside allied German units, the latter indicating the continued relevance of the idea of the Holy Roman Empire, which compromised modern Austria, Germany and the Czech Republic. In the battle zone, fortifications were improved, while local sources of supply were depleted, and fighting in marshy, fever-ridden lowlands was difficult. The Austrians recaptured Grosswardein in Transylvania in 1692, but in 1693 were forced to raise the siege of Belgrade. Matters came to a climax as a result of the accession of the energetic Mustafa II in 1695 and the end of the war with France in Italy in 1696, which enabled Leopold I to transfer more troops and his rising general, Eugene of Savoy, to Hungary. Mustafa had some success, storming Lippa and relieving Temesvár in 1695, and outmanoeuvring the imperial army in 1696, but, in 1697, Eugene moved rapidly and attacked Mustafa at Zenta as his army was crossing the river Tisza. The Turks, caught unprepared and divided by the river, lost both troops and supplies heavily. The victory made Eugene's reputation. His forces went on to raid Bosnia, sacking Sarajevo, but it was too late in the year to attack Temesvár or Belgrade. In the following year, the Turkish forces refused to engage, while Eugene's army was troubled by a mutiny caused by lack of funds. Leopold wanted peace in order to concentrate on the challenge of the Spanish succession, the contentious nature of the succession to the childless Charles II of Spain. Under the Peace of Karlowitz (1699), Austria kept Transylvania, and Hungary bar the Banat of Temesvár, a major territorial shift. Venice gained the Morea and a strip of Dalmatia, while Poland acquired Podolia.

War resumed in 1716, when the Austrians became concerned at Turkish gains at the expense of Venice: the Morea was conquered in 1715 in one of the most decisive campaigns of the century and one that consisted of sieges with no battle. The Turks refused to withdraw, instead hoping to reconquer Hungary. In the opening campaign, the imperial army under Eugene defeated the Turks under the Grand Vizier, Damad Ali Pasha, at Pétervárad (Peterwarden) near Belgrade on 5 August 1716: the Turkish janissaries had some success against the Austrian infantry, but the Austrian cavalry drove their opponents from the field, leading to the defeat of the janissaries. The Grand Vizier was also killed. Eugene pressed on to attack Temesvár, a major fortress which had defied the Austrians during the previous war, and which controlled or threatened much of eastern Hungary. It surrendered after heavy bombardment. The following June, Eugene crossed the Danube using pontoon bridges and besieged Belgrade, the key fortress in the region. The Grand Vizier brought up a relief army and began bombarding the Austrians from higher ground. In a difficult position, Eugene resolved on a surprise attack, and, on the morning of 16 August, advanced through the fog to defeat the Turkish army, capturing all its artillery. This led to the surrender of Belgrade six days later. In 1718, by the Peace of Passarowitz, Austria gained the Banat of Temesvár, Little (western) Wallachia, and northern Serbia, including Belgrade.

Russia

The Russians were less successful against the Turks in this period. Azov, a peripheral, but important, position, was captured in 1696 after an initial attempt the previous year had failed; but when Peter the Great moved closer to the centre of Turkish power and invaded the Balkans in 1711, he was outmanoeuvred at the river Pruth. The much larger Turkish forces had moved faster than anticipated and surrounded the Russians, and their superior artillery bombarded the Russian camp at Nou Stănileş ti. Turkish attacks were repelled only with difficulty, and, short of food, water and forage, Peter was forced to sign a peace treaty, in which he made concessions.

The contrast between this failure and Peter's much greater success against the Swedes should not automatically be taken as an indicator that the Turks were more effective than both of these armies, not only because it is always problematic to generalise from individual episodes, but also because Russia did not show the same commitment it devoted against the Swedes at the expense of the Turks until later in the century. Comparison is made even more difficult if it is extended to consider Chinese success over Galdan of the Zunghars at Jao Modo in Mongolia in 1696. In some respects, this was a similar campaign to that of Peter, but Chinese success does not prove an inherent superiority to Russian war-making. Instead, both Russia and China faced serious logistical problems, and this dimension was crucial in their campaigning.[1]

Peter I (the Great)'s military reforms were important, although they had been prefigured under his predecessors, and there was a considerable degree of continuity.[2] Conscious of Swedish developments under Gustavus Adolphus, who blocked Russia from the Baltic in a war in 1611–17, Tsar Michael (r. 1613–45), Peter's grandfather, decided in 1630 to form 'new order' military units, officered mostly by foreigners. In 1634, however, the new units were demobilised and the foreign mercenaries ordered to leave. Tsar Alexis (r. 1645–76) recruited foreign officers and armed and organised troops on the Western model, but, in peacetime, the army was run down and poorly trained. Failure at Narva (1700) encouraged Peter to press ahead with his policy of strengthening the army. Relying on the principle of conscription, Peter introduced in 1705 a system of general levies based on a Swedish model. Every twenty taxable households were ordered to send one recruit, and they were made responsible for his replacement if killed or incapacitated. This raised over 150,000 men in 1705–9. New regiments were created, twelve in 1705–7 alone, and by 1707 the army was about 200,000 strong. The practice of recruiting foreign officers continued. In 1704, the Government took over responsibility for feeding and clothing troops. Peter also developed Russia's armaments industry and introduced uniforms and new weaponry. The old Russian flintlock musket was officially supplanted by the Model 1709, an up-to-date design which did not weigh too much and was fitted with a socket-type bayonet, although the supply of the gun was irregular. Foreign experts were used in producing new weapons. For example, a Dutch artillery specialist, de Konning, became superintendent of foundries in Karelia and played an important role in the production of cannon for the new Baltic navy. The War College, established in 1718–19, brought a measure of coherence to military administration, although this remained limited.

Peter's military reforms extended to the navy. A fleet was constructed for the campaign that led to the capture of Azov from the Turks in 1696, foreign experts were imported, and Russians were sent abroad to learn shipbuilding. The subsequent development of the fleet reflected Peter's personal enthusiasm. This led him to devote much time to the details and to study shipbuilding in the United Provinces and Britain. The development of Russian naval power also reflected a clear goal in the shape of the need to challenge Sweden for control of the Baltic if Russia was to succeed in dislocating the structure of the Swedish empire and preventing a Swedish reconquest of her eastern Baltic provinces, which had been lost to Russia in 1710, a clear instance of the changing nature of strategic culture – in this case the shift from a wish for a Baltic coastline to a drive for power in the Baltic. A naval academy and a large artillery yard were constructed at St Petersburg and a school of navigation at Moscow. By the time of Peter's death in 1725, Russia had a fleet of thirty-four ships of the line and numerous galleys, and the raids the latter mounted on the Swedish coast played a major role in forcing Sweden to make peace in 1721.

Education was an important theme in Peter's reforms. He founded artillery (1701) and engineering (1709) schools, which graduated 300–400 officers annually, while officers were trained by service as ordinary soldiers in the guards regiments. Peter insisted on progressing through the ranks himself and tried to ensure that no noble received a commission without some form of training. His attempt to make state service a central focus for the aspiration of many led to a table of ranks designed to link status with service and to the spread of uniforms, themselves a mark both of service and of the state's role in allocating rank. Military uniforms became the principal dress of the nobility, while Peter was the first monarch to require all Russian soldiers to wear specified uniforms. Under Peter, the armed forces, particularly the army, replaced the church, with which his relations were poor, as the lodestar of monarchical action and, to an extent, of national unity. Instead of being a semi-sacral figure, Peter made the monarch a military leader, and a professional in this role. The nature of his successors from 1725 to 1796 – four women, one youth and only one adult male, Peter III (who reigned for less than a year) – limited the furtherance of this process, but there was no return to the earlier image.

In part, Peter's policy was driven by the requirements of the war with Sweden (1700–21), while the *streltsy* revolt of 1698 had underlined the need for a reliable military. Initially, the war with Sweden, which Peter began in alliance with Augustus II of Saxony-Poland and Frederick IV of Denmark, went badly. A Swedish landing in Zealand led Frederick to abandon his allies, while Peter was soon after defeated at Narva (20 November 1700). The Russians had besieged the Swedish fortress of Narva, only to face a relieving Swedish army under Charles XII. The latter's ability to seize and use the initiative was crucial, not least because the Russian army was far larger. The Swedes advanced rapidly and broke into the Russian entrenchments in two columns. The Swedes rapidly came to hand-to-hand combat and proved adept with their bayonets. A snowstorm blew directly into the faces of the defenders, and the Russian position collapsed, with their troops fleeing or surrendering. The Swedes lost 2,000 dead and wounded, the Russians 8,000–10,000 and all their artillery. As with Vienna in 1683 and Turin in 1706, Narva demonstrated the vulnerability of a poorly commanded and badly deployed siege army to a relief attempt. The attackers in each case were able to concentrate mass and break through the defending force.

Swedish involvement in Poland over the following years, where Charles won battles but found it impossible to create a lasting settlement,[3] allowed Peter to increase the size and improve the fighting capacity of his army. In 1701–3, the Russians scored successes against the Swedes and in 1703 Peter founded St Petersburg. In 1704, the Russians successfully besieged Narva and Dorpat; the Swedish forces that tried to relieve them were short of supplies and outnumbered. In late 1707, Charles turned from Poland and launched his campaign against Russia. He successfully crossed the river lines that blocked his advance, but the Russian scorched-earth policy created grave difficulties. Supply problems, the severity of the winter of 1708–9, and the hope that Mazepa, Hetman (elected leader) of Ukraine, would raise his people against Peter, led Charles to turn southwards. However, Mazepa's inadequate response and the savage cold of the winter undermined Charles' move.

The gruelling Swedish campaign came to a disastrous end at the decisive Russian victory of Poltava (17 June 1709). This was no second Narva and showed the very different results of the same policy, an attack on an entrenched Russian position. The Russian army was entrenched behind redoubts, and the Swedes did not use their artillery to try to weaken the Russian position, probably because, short of supplies, Charles preferred to rely on a rapid infantry attack, rather than on the attrition of a firepower exchange. In contrast, the Russian commander, Peter the Great, placed great weight on artillery. Whereas, at Poltava, the Swedes had only four cannon, the Russians had 102, twenty-one of them heavy pieces, and plentiful ammunition; their cannon fired 1,471 shot. The Swedes penetrated the Russian redoubts at heavy cost, but, in the final engagement, the Russians advanced from their camp with their entire force, and their superior strength and firepower proved decisive.

The Russians lost about 1,300 dead, the Swedes 6,900, as well as 2,800 prisoners, a high ratio from an army of 19,700. The defeat turned into disaster when most of the retreating army surrendered to their Russian pursuers three days later.

Charles took refuge in the Turkish empire, where he stayed until 1714, but, in the meantime, the Swedish position around the eastern Baltic collapsed. In October 1709, the Russians occupied Courland, and in 1710 successfully besieged Viborg, Tallinn, Mitau and Riga, following with the conquest of Finland in 1713–14. In 1714, the Russian fleet defeated the Swedes off Cape Hangö. In 1716, Peter moved troops into northern Germany and prepared an amphibious invasion of Scania (southern Sweden) from Jutland, a major shift in strategic goal and operational means. No such invasion was launched, and Peter was obliged to withdraw from Germany, but a British-backed attempt in 1720 to create a large coalition in order to intimidate him into returning his conquests failed, and the Peace of Nystad (1721) left him with Estonia, Ingria, Kexholm and Livonia.

Russian military strength and its potential impact was thereafter to be a theme of major concern for other states. Apart from brief interludes in 1918–19 and 1941–3, this was a fundamental change in Western geopolitics. Russia achieved this situation without any qualitative military advance on the remainder of Europe: Russian weapons, troops, warships and commanders were not superior. Instead, this was a case of successful Westernisation combined with the latent strengths of Russia as a military power, albeit strengths that had earlier been affected by political problems and economic and fiscal limitations.[4] A substantial population acted as a solid base for recruitment and taxation. As a result, Russia had a far larger army than either Sweden or Prussia, despite these having a higher percentage of their male population under arms. The metallurgical riches of the country were exploited under Peter, so that the production of handguns rose from 6,000 in 1701 to 40,000 in 1711, and, as the largest producer of iron in Europe, Russia necessarily played a major role in what was still an iron age. New foundries were established in the Urals and Siberia, and, while the quantity of Russian artillery increased, the cannon were also organised into categories and uniform standards of calibre set in 1706. When applied to military ends by a determined moderniser, of which Peter was the obvious example, the consequences of Russian resources were a substantial growth in power and in the ability to apply it.

The nature of weapons (and other) technology, and the Western focus of the Russian state were such, however, that Russia's direct global impact was limited. This was particularly because the development of the navy under Peter was not continued by his immediate successors and Russia did not, from her main naval bases, create a colonial empire. The size of Siberia and the impossibility of sailing along its Arctic coastline were such that exploration and settlement on the Pacific coast of North America were peripheral to the centres of Russian power. Russian force projection was demonstrated in 1770 when the fleet that had sailed from the Baltic to the Mediterranean the previous year won a spectacular naval victory over the Turks at Cesmé. The Russians, however, lacked a naval reach capability comparable to that of the western European powers. The absence of warm-water colonies was important, for it deprived the Russians of the bases necessary for further activity. Russia was essentially a land power, and her society had little interest in maritime activity. Instead, as in the seventeenth century, it was on the Euro-Asian land frontier, especially that with Turkey, that Russia sought to expand. Although Peter the Great was not particularly successful in this sphere, the balance of advantage was increasingly with Russia. Thanks to Peter, the distribution of military power within Eurasia had shifted. China remained the strongest land power, but Russia was now the strongest non-Eastern state.

The wars of Louis XIV

This provides a context, and sense of relative importance, within which to consider the topic that usually engages the bulk of scholarly attention, the wars of Louis XIV. Soon after the start of the Nine Years' War (1688–97), Louis expressed his sense of confidence, 'Whatever happens in this war, it is certain that the good state of my frontiers and of my troops will prevent my enemies from troubling the peace of my kingdom and will give me the means to extend my possessions.'[5] Triumphs, such as Louis's captures of Mons (1691) and Namur (1692), were celebrated with services and paintings of commemoration.[6]

In fact, the war showed how Louis was unable to determine outcomes. What was intended as a small-scale conflict – in 1688, French forces advanced into Germany in order to press Louis's claims in the Rhineland – became a major struggle. The conflict rapidly widened, and, by the summer of 1689, Louis was at war with Austria, Bavaria, Britain, the Netherlands and Spain. He had declared war on the last because he saw it as vulnerable; in 1690, Victor Amadeus II of Savoy-Piedmont joined the alliance against Louis. Louis's isolation was a significant diplomatic defeat and greatly influenced the course of the conflict. In contrast, the French had allies in the Spanish (1701–14), Polish (1733–5), and Austrian (1741–8) Wars of Succession, and in the Seven Years' War (1756–63).

The diversion of much of the Anglo-Dutch military effort to the conquest of Ireland in 1690–1 gave Louis an opportunity, while the financial weakness of the Spanish administration in the Spanish Netherlands seriously exacerbated poor relations in the Allied coalition.[7] The aggressive François, Duke of Luxembourg invaded the Spanish Netherlands in 1690 and defeated George Frederick of Waldeck's army at Fleurus (1 July), by turning its left flank while mounting a frontal assault. Waldeck was forced to create a new front with his reserves and second line, and, attacked from two directions, had to make a fighting retreat. The French retained the initiative in 1691, successfully besieging Mons and, in a relatively minor engagement, defeated

Waldeck at Leuze (19 September) by a surprise attack. On the Italian front, the French captured Nice and Montmélian from Victor Amadeus II. In 1692, Luxembourg captured Namur, a siege rapidly driven forward to Vauban, helped by a siege train of 151 cannon: the fortress fell after five weeks. Luxembourg also narrowly defeated William at Steenkirk (3 August), following in 1693 with victory at Neerwinden (29 July) after a hard struggle that involved repeated French assaults on the entrenched Allied position. However, the commitment of Anglo-Dutch strength denied Louis decisive victory, and the loss of Namur in 1695 shook French prestige. Dutch logistics were better than in the 1670s. Also in 1695, Allied forces successfully besieged Casale in north Italy in a major demonstration of Spanish resilience.[8] France was also put under pressure from the harvest famine of 1693, which hit tax revenues as well as food supplies. This was significant because the war had become attritional, with Louis unable to maintain an operational momentum on any one front.[9]

The anti-French alliance collapsed in 1696, with the defection of Victor Amadeus, but Louis only gained this by concession, particularly the cession of the fortress of Pinerolo, as well as the return of French gains; this reflected the failure of French operations in Italy and is a comment on the combination of France's international failure with the limited capability advantages that stemmed from its military developments. In 1697, moreover, peace negotiations affected the continuing course of the conflict, as they so often did in the wars of the period. In the Spanish Netherlands, Vauban successfully besieged Ath, while the French capture of Barcelona in 1697, in which costly assaults took precedence over 'scientific siegecraft', leading to criticism by Vauban, led Spain to sue for peace. By the Treaty of Rijswijk (1697), Louis had to accept some territorial losses, including the fortresses of Breisach, Freiburg and Luxembourg, but he prepared the ground for the diplomacy round the Spanish succession.

The death of the childless Charles II of Spain in 1700, and his will in favour of an undivided inheritance by Philip, Duke of Anjou, brought Louis's second grandson to the Spanish throne, without the partition of the inheritance negotiated in earlier treaties of 1698 and 1700 between Louis and William III; but Philip was challenged by a rival claimant, Leopold I's second son, Charles. War broke out in 1701 and, in the following year, Britain and the Dutch joined Leopold's side, William's legacy to his successors. As with Napoleon's later repeated political failure in this handling of other powers, in part the creation of an anti-French coalition reflected a series of provocative French steps that alienated opinion in Britain and the United Provinces and made it much easier for William to commit them to the anti-French camp.

In Germany, the Archbishopric of Cologne, allied to France, was overrun by the Allies in 1702–3, which strengthened the Allied position in the Low Countries, but the combination of French and Bavarian troops in southern Germany threatened the collapse of Austria. Leopold, however, was saved in

1704 by the Blenheim campaign: the march to the Danube of an Anglo-Dutch army under John Churchill, 1st Duke of Marlborough, his juncture with the Austrian forces and his major victory at Blenheim (13 August). Attacks on the French flanks led to the committal of their reserves, and then Marlborough broke through the French centre. This victory led to the Allied overrunning of Bavaria, a conquest retained during the war. After an inconclusive campaign in Flanders in 1705, Marlborough again won victory at Ramillies in the Spanish Netherlands (23 May 1706) by breaking the French centre after French troops had been deployed to support action on a flank. He went on to conquer a number of weakly fortified positions, but the problems of exploiting victory in the face of ably defended and well-fortified positions was shown by the remainder of the campaign.[10]

The French attempt to regain their position in the Spanish Netherlands was thwarted at Oudenaarde (11 August 1708), where, after several hours fighting, they were almost enveloped by Marlborough's cavalry. The French lost heavily and Marlborough exploited the victory to besiege and capture Lille. Vauban's fortifications, nevertheless, proved their value as the Allied forces invaded France. Whereas most of the fortresses in the Spanish Netherlands had fallen rapidly to Allied attack, Lille held out for 120 days in 1708 before capitulating on 9 December, too late for the Allies to make further advances before the onset of winter. A poorly coordinated attack on defensive positions on 7 September had left nearly 3,000 attackers dead or wounded, and the Allies were only successful when they concentrated their artillery fire.

Marlborough was less successful the following year at Malplaquet (11 September), as his tactics had become stereotyped, allowing the French to devise a response, as the Austrians were later to do to Frederick the Great of Prussia. The French held Marlborough's attacks on their flanks and retained a substantial reserve to meet his final central push. The French eventually retreated, but their army had not been routed. The casualties were very heavy on both sides, including a quarter of the Anglo-Dutch force: the battle was the bloodiest in Europe before the Franco-Russian clash at Borodino (1812). Malplaquet was followed by the capture of Ghent, but hopes of breaching the French frontier defences and marching on Paris were misplaced. In 1711, Marlborough captured the fortress of Bouchain, but, in light of a change in the political situation within Britain, and a disenchantment with the lengthy land war, such achievements could no longer keep Britain in the conflict.

This was part of a more general failure to invade France successfully, a failure that reflected both Alliance weaknesses and French strengths. Marlborough's plan to invade Lorraine up the Moselle valley in 1705 had to be abandoned due to lack of German support. The Austrian invasions of Alsace in 1706 and the Franche-Comté in 1708, and a combined Anglo-Austrian-Savoyard attack on Toulon in 1707, were also unsuccessful. The last was intended to seize the French navy's Mediterranean base and to expose southern France to attack. Austrian and Sardinian forces advanced overland, while the British fleet provided naval support. The French, however, were able to send reinforcements, and the siege was acrimoniously abandoned, although it served to confirm France's weakness at sea.

The French, moreover, were driven from Italy. In the culmination of campaigning that had begun in 1701, and which had seen the French overrunning most of Savoy-Piedmont, their army besieging Turin was defeated by Eugene and Victor Amadeus (7 September 1706). This major triumph tends to be underrated because of the focus on Marlborough's victories and the tendency to downplay the importance of Italy. The battle also had a strategic outcome: the French withdrew from Italy and the way was therefore clear from the Spaniards to lose Naples to an Austrian invasion in 1707.

The Grand Alliance had far less success in Spain. Despite the support of Catalonia and Valencia, and the intervention of British, German and Portuguese troops and naval power, the attempt to establish Archduke Charles as Charles III failed. Thanks to British naval strength, amphibious forces captured Gibraltar in 1704, Barcelona in 1705 and Minorca in 1708, while Madrid was occupied briefly in 1706 and 1710, but Castilian loyalty to Philip V and Louis XIV's support proved too strong. Philip's cause became identified with national independence, despite his reliance on French troops. They defeated the Allies at Almanza (25 April 1707), forcing Charles back to Catalonia, although it was not possible to exploit this in order to end the war in Spain. In 1710, Charles defeated Philip at Almendra (28 July) and Saragossa (19 August), before occupying Madrid, but few Castilians rallied to him and his communications became hazardous. As a result, he withdrew from Madrid and, at Brihuega (9 December), part of his retreating force was attacked by a superior army under Vendôme and forced to surrender. On the following day, another section of his retreating army fought off a French attack at Villaviciosa, but Charles had now lost Castile. His Catalan supporters continued to resist, but they were defeated in a series of sieges: Gerona fell in 1711 and Barcelona in 1714. Majorca was captured in 1715 once the cover of British naval protection was removed.

Naval strength had made British intervention in Iberia possible, as well as enabling the British to inhibit French invasion planning and to maintain control of maritime routes to the Low Countries. The improvement in British naval capability was demonstrated with the development of revictualling at sea in order to support a fleet maintaining watch on Brest, although manning the navy remained a serious problem. Naval strength also secured for Britain its substantial trade surplus, and this, in turn, provided the financial surplus that permitted the payment of subsidies to Savoy-Piedmont, Denmark, Hesse-Cassel, Prussia, Austria, Portugal, Saxony and Trier, and thus helped maintain opposition to France on land in Europe.

The war encouraged and interacted with a number of rebellions, most significantly those against Philip V in Spain, but the latter were defeated, while the pro-Jacobite French invasion attempt on Scotland in 1708 was thwarted by the British navy, the Camisard (Huguenot) rising in the Cévennes in south France was suppressed, and the Hungarian Rakoczi rising against Habsburg rule in 1703–11 led to an eventual settlement. These failures did not demonstrate the frequently proclaimed monopolisation of force by government, still less the demilitarisation of civic society, but they showed the general direction of military power towards the state. This was also to be further seen with the defeat of the Jacobite rising in Scotland and northern England in 1715–16.[11]

The course of the War of the Spanish Succession reflected the security of the French home base and the French ability to campaign simultaneously on several fronts, but France's opponents were also effective, and, albeit with considerable difficulty, developed considerable experience with combined (allied) and joint (army and navy) operations. Indeed, participation in the alliance helped maintain the military effectiveness of second-rank powers, such as Prussia and Savoy-Piedmont, as well as their lesser counterparts, for they benefited from subsidies from their wealthier allies, particularly Britain.[12]

The campaigning showed that the French generals were not without talent, Marshal Villars being particularly impressive. He defeated the Allies (no longer including the British) at Denain in north-east France on 24 July 1712, and, to the east, captured the fortresses of Freiburg, Kehl and Landau in 1713. However, other French generals were outmanoeuvred and outfought by Marlborough and Eugene. Furthermore, the Nine Years' War and the War of the Spanish Succession placed major strains on the system of French army supplies. Much of the fighting was on or near French territory, which became exhausted because the conflicts lasted for many years, while economic circumstances anyway were generally harsh. There were years of poor harvests and higher-than-normal mortality rates, which affected both the supply of food and men. The militia established in 1688 for garrison duty within France was used to replace casualties in the regular army. In 1709–10, a period of especially harsh conditions, the system of supplying the army through negotiating contracts with army suppliers broke down, as the government could no longer afford it, and it was instead obliged to experiment by providing supplies through its own officials. More generally, shifts in the international system and the enhanced proficiency of the Austrian and British armies were such that France was less capable of projecting its power than had seemed to be the case in Louis's early decades, and this impression was strengthened when the French army was drastically cut after the end of the War of the Spanish Succession (1714) and the death of Louis XIV (1715).[13]

1714–21

The years after the war indicated the continued variety of European conflict. Aside from the continuing Great Northern War (1700–21), and conflict in the Balkans, including the unsuccessful Turkish siege of Corfu in 1717, there was an insurrection in Britain: the unsuccessful Jacobite uprising of 1715–16, which involved two full-scale battles. Furthermore, Britain, France, Savoy-Piedmont, Austria and Spain were involved in the War of the Quadruple Alliance (1718–20), which arose from the determination of Philip V to regain the territories Spain had lost in Italy. Although frequently poorly paid, the total size of the Spanish army rose, and a major effort was put into expanding the size of the navy. The conflict involved most of the powers of western Europe, as Spain was opposed by the defenders of the 1713–14 peace settlement, but also demonstrated the primacy of politics. Aware of the unpopularity within France of war with Spain, and also unwilling to destroy Spanish power, the French launched only a limited invasion of northern Spain in 1719. Yet more than politics was involved: two military successes in 1718 were very important to the course of the conflict: the Spaniards were able to land 20,000 troops on Sicily and to overrun much of the island, but the British defeat of a poorly deployed Spanish fleet off Cape Passaro enabled the Austrians to counter-attack, leading to battles, such as Francavilla, which are generally ignored. The range of problems affecting amphibious operations was shown in 1719 when a Spanish invasion of Britain failed, with most of the force dispersed by storms: the small force that landed and the supporting Jacobites were defeated in Glenshiel. The war ended with Spain, without allies, having to accept Austrian control of Sicily.

Colonial conflict

As with other wars between Atlantic European powers from the mid-sixteenth to the mid-eighteenth centuries, the conflict had a transoceanic dimension, but, prior to the Seven Years' War (1756–63), this dimension was not crucial to the struggle between the states, with the exception of the mid-seventeenth century struggle between Portugal and the Dutch. The French navy proved particularly neglectful of the New World, ensuring that French capability there became largely a matter of activity on land.[14] However, thanks to the failure (as a result of the running aground of part of the force in the St Lawrence) of the British amphibious expedition against Quebec in 1711, British pressure on New France was, prior to 1758, largely peripheral, focusing as it did on Nova Scotia, Cape Breton Island and, to a lesser extent, Hudson Bay, and, instead, the Iroquois proved the threat to the core of the French colony.

Across the Atlantic, the War of the Quadruple Alliance interacted with tension over French expansion from its already-established colony in Louisiana, which threatened the Spanish position in Florida and Texas. In 1719, the French captured Pensacola, the major Spanish base in West Florida, in a surprise attack, but it was recaptured by an expedition of 1,400 troops from their base of Havana, only to be retaken by the French and their local allies, the Choctaws. Pensacola was returned to Spain when the two sides made peace. The Spaniards had also sent a force from Santa Fé in 1720 to counter the French expansion, but most of it was wiped out by Native Americans on the South Platte.

Conclusions

The repeated wars of the period helped underline the value of effective forces and also led to an increase in their size. The wars led to a new international hierarchy, in which Austria, Britain and Russia were more prominent, while France and, even more, Spain and Sweden were less so. The Dutch played a key role in the victorious coalition that triumphed in the War of the Spanish Succession, but financial retrenchment helped lead to a major fall in the size of the army from a peak of 119,000 in 1708 to 34,000 in 1716–17. Britain, instead, became increasingly dominant in relations with France. On the global scale, the British strengthened their position in North America, at the expense of both France and Native Americans;[15] but it was the more secure position of Britain and Russia within Europe, and their development of military systems reflecting economic and governmental resources, that were of greatest long-term significance.

6

WARFARE, 1722–55

Compared to the period that followed, particularly the Seven Years' War (1756–63), the years 1722–55 have received insufficient attention, in large part because the wars appear less important: they have certainly featured little in the collective myths of subsequent ages. For example, the French army was effective both during the War of the Polish Succession (1733–5) and the War of the Austrian Succession (1740–8, the French taking part from 1741), but there has been relatively little research on this period, especially on the operational dimension. As a consequence, work on Frederick the Great's Prussia,[1] which focuses on the Seven Years' War, lacks an adequate comparative dimension. Victory over the French at Rossbach in 1757 (see p. 95) is employed to assert a systemic Prussian advantage that is misleading, not least for the 1740s when French forces in the Low Countries were commanded by Marshal Saxe, an outstanding operational commander, as well as an imaginative thinker on the nature of war. Maurice, Count of Saxe's generalship, was instructive not only because of the battlefield ability he displayed in 1745–7 to control large numbers effectively, in both attack and defence, but also because of his determined espousal of a war of manoeuvre. His preference for bold manoeuvres, emphasis on gaining and holding the initiative, and stress on morale, contrast markedly with stereotypical views of non-Frederician mid-eighteenth century warfare, as indeed do the Russian invasions of Moldavia and Finland in 1739 and 1742 respectively. Work on Prussia could profitably be integrated with that on other forces in order to gain a better grasp of the extent and limitations of Prussia's comparative advantages, which otherwise tend to be seen in isolation.

As far as conflict with non-Western forces is concerned, there is also a tendency to downplay the period in favour of apparently more spectacular wars before and after. Austrian failure in war with Turkey in 1737–9 attracts less attention than success in 1683–99 and 1716–18, while Russian success in 1735–9 is downplayed in favour of more dramatic and consequential victories in the wars of 1768–74 and 1787–92. In India, war from 1756, particularly British victory at Plassey in 1757, similarly seems more significant than what came earlier, although the latter indicates the difficulties that could face

Western forces. Thus, the Portuguese were hard-pressed in west India in 1737–40 when they were involved in a disastrous war with the Marathas. Their positions of Salsette, Bassein and Chaul were captured, and they very nearly lost their major base, Goa. As a consequence of this, and of the general decline of Portuguese power, the British, based in Bombay, a dowry that came with Charles II's wife, became the major Western force in west India.

The Dutch remained the leading Western power in the East Indies (modern Indonesia), although the Dutch East India Company was far weaker than its British counterpart (which focused on India), and the Dutch faced serious problems in Java. There, the Dutch company's interventions in the persistent civil wars in the kingdom of Mataram were weakened by the inability of the company's army to operate successfully away from the coastal areas, not least due to an absence of naval support. More generally, the effectiveness of the Dutch in Java depended on local allies, Cakraningrat IV of Madura saving their situation in 1741 when they were hard-pressed. The balance of military and political advantage was always shifting, and any unexpected pressure could lead to crisis. As a consequence, in the Third Javanese War of Succession (1746–57), the Dutch suffered defeats in 1750 and 1755, while, in the kingdom of Banten in west Java, a rebellion in 1750 led to the defeat of Dutch forces. Further afield, on the island of Sulawesi, a Dutch attempt in 1739–40 to crush the dynamic Arung Singkiang, ruler of Wajo, had only limited success: disease and bad weather greatly hindered the Dutch.

Knowledge and theory

Aside from the specific conflicts of the period, it is instructive to consider longer-term trends. One that has been discerned is the development of scientific methods. Cultural and intellectual factors acted as enablers for change and development, not least in terms of perceiving problems and constructing solutions. These factors can be summarised in terms of science, with reference made to the scientific revolution, but these terms are only appropriate if it is appreciated that they should not be seen in terms of the abstractions of much modern science, but rather as a method of rational analysis that was applied with increasing effectiveness, particularly in the seventeenth century. Mathematics was applied with great effect in ballistics and navigation, and a 'knowledge nexus' developed that was of particular value in the projection of European power, as cartography and navigation were employed to gain conceptual understanding of the world.

Knowledge was also a matter of the analysis of classical texts in order to establish general principles of war-making. Thus, in the 1720s, the French veteran Jean-Charles, Chevalier de Folard focused his work, which sought to replace firepower by an emphasis on shock in the shape of attacking infantry columns, on a lengthy commentary on the Roman author Polybius, specifically on the latter's account of the clash between Macedonian phalanx and

Roman legion. Folard wanted to bring back the wedge (*cuneus*), while the column he advocated was a form of the phalanx.[2] Saxe, the leading French general of the 1740s, called his ideal formation a legion as a classical affectation, and was interested in the reintroduction of armour. Folard and Saxe were also interested in reviving the pike, as was the *Encyclopédie* (1751–65). In 1736, the British envoy in Berlin reported, 'They make frequent parallels here between the Macedonian troops and theirs.'[3]

Referring back was not the same as learning from the past. Nevertheless, the process of validation was one in which the balance between modernity and history was very different to the situation today. The long-standing reverential and referential hold of the fourth- or fifth-century AD *Epitoma rei militaris* (*On Military Matters*) by Flavius Vegetius, was such as to make historical study difficult in the sense of detecting, describing and analysing change. Folard found it normal to debate with Vegetius as if he were a contemporary, while Franz Miller, an influential military theorist and the author of *Reine Taktik der Infanterie, Cavallerie und Artillerie* (Stuttgart, 1787–8), regarded the Roman military as if it were an army of his own time which he sought to improve. However, the response to antiquity was flexible, with Saxe, for example, in his *Rêveries sur l'art de la guerre* (1732) drawing on Vegetius, but also offering a different tactical tradecraft that, in part, reflected the potential of firepower.[4]

More generally, there was an emphasis on scientific process as a sign of professionalism and of the rational methods of control that this was supposed to entail. This was seen, for example, in the signalling systems advocated in French naval manuals as a way to enhance command and control.[5] Scientific methods entailed not only the concern of generals with sieges and guns, but also the use of scientific knowledge at the operational level, with the need to plan foraging and marches requiring an understanding of agronomy, surveying, celestial navigation, botany and forestry. Military technology and practice were thus influenced by the larger economy of knowledge, which expanded considerably in this period, helped by the diffusion of information through the culture of print.

This can be seen in the case of Austria, with the groundwork of the drill and artillery reforms of the 1750s being laid two decades earlier; and the same was true of the foundation of the Austrian military academy. Thus the 1730s and 1740s come out of the shadows, as they also should do for other militaries. This argument has been anchored by a detailed consideration of the Austro-Turkish war of 1737–9, a conflict that had earlier received insufficient attention. During that war, commanders with staff-planning skills were at a particular premium, and this led to a demand for intellectual accomplishment and technical skill. At the same time, the sometimes almost abstract treatment of military development needs to be given a contextual understanding in terms of divisions over goals, methods and influence. Generals were participants in the continual struggle of military politics, and, in the case of this war, the Austrian generals were free to choose their own particular approaches to the campaigning based on which interest their ambition allied them with.[6] This can be paralleled with other militaries, and is a reminder of the need to anatomise strategic cultures in terms of these politics.

The wider world

There has been insufficient research to evaluate the extent to which the application of knowledge for military purposes in the West was also seen elsewhere. In part, this is because it is not only in the West that these are relatively obscure years. Most obviously, there has been far too little written on the dramatic and far-ranging struggles in South Asia. In 1721–2, the Ghilzais of southern Afghanistan overthrew Safavid Persia, touching off a period of serious instability in Persia that led to the seizure of control in 1730 by Nadir Shah. He embarked on a programme of expansion in all directions that saw Persian forces capture Kandahar (1738), Delhi (1739), Bukhara (1740) and Khiva (1740), leading to a range and personalisation of military activity that prefigured that of Napoleon, with similarly short-term military consequences. Nadir Shah's major conflict was with Turkey, but, although the Turks were driven from northern and western Persia, which they had occupied in the 1720s, it proved impossible to capture Baghdad and Mosul. Conflict between Indian rulers in this period, as Mughal control weakened, also tends to be neglected.

In both spheres, there is, instead, an emphasis on Western intervention. In fact, this was still relatively peripheral in Persia. Peter the Great had taken advantage of Persian weakness to advance south along the western shores of the Caspian Sea, capturing Derbent (1722), Baku (1723) and Rasht (1723), but the conquests to the south of the Caspian were abandoned in 1732: the Russians found them of little value and lost large numbers of troops to disease. In India, the British and French were more significant in the Carnatic in the 1740s and early 1750s, taking opposite sides in a succession struggle, but, as yet, the British were not as successful as they were to be in Indian power politics in the 1760s and, even more, 1790s–1800s. Indeed, the key development in the early 1750s was Afghan pressure, with the overrunning of Punjab and Kashmir in 1752. This looked towards the major Afghan victory over the Marathas in the Third Battle of Panipat outside Delhi in 1761.

Europe, 1722–35

The period 1722–32 was a period of apparently imminent war within the West, but of singularly little conflict. An ineffective Spanish siege of British-held Gibraltar in 1727 was without consequence in what proved to be a short semi-war between Britain and Spain (1726–7) that did not develop into full-scale hostilities. Furthermore, fears and plans, for example about a Russian attack on Sweden (1723), an Austro-Russian invasion of Hanover (1726), a

Prussian invasion of Hanover (1729), a British conflict with Spain (1729), an Anglo-French-Spanish attack on Austrian Italy (1730), a British invasion of France and vice versa (1731) and a Spanish invasion of Austrian Italy in 1732 were all unrealised. War did not break out until 1733, but earlier concerns and preparations indicated the extent to which conflict was seen as an ever-present possibility. The prospect of conflict helped drive alliance politics, with states being assessed from what their forces could bring in any war. Thus, in 1723, Charles, 2nd Viscount Townshend, the key minister in British foreign policy trumpeted the value of a recently concluded alliance with Frederick William I of Prussia,

> in effect puts into the scale with His Majesty [George I] the whole force and strength of Prussia, at least three score thousand men, excellent troops. Before this the power of Great Britain lay only in its fleet, which though strong, and of great command in maritime cases; yet . . . the respect our fleet carried could not spread its influence so far as was necessary. But now . . . His Majesty is become master as it were of so mighty a land force, he will not only be more secure, but also more respected both in the North and the South, and have it in his power to act more independently from the houses both of Austria and Bourbon.[7]

In practice, fearful of Austria and Russia, Frederick William was to abandon George in 1726.

The War of the Polish Succession (1733–5), a conflict touched off by a contested election for the crown of Poland, which was used as an opportunity by France and her allies, Spain and Sardinia, to attack Austria, demonstrated that, alongside the usual contrast between operations in western and eastern Europe, in terms of the far greater mobility of the latter, there was also much variety in both spheres. In the Rhineland, the successful French sieges of the fortress of Kehl (1733) and Philippsburg (1734) did not lead to any decisive military verdict, there were no battles, and feared French moves into northern Germany were not made: the successful French advance down the Moselle valley in early 1734 was not followed up because of the crucial political context. The French government was urged by the Bavarian government to occupy Cologne and advance via Germany into Bohemia, but it did not wish to bring neutral powers, particularly Britain and the United Provinces, into the war. This also led the French, by a convention with the Dutch in 1733, to agree not to attack the vulnerable Austrian Netherlands (Belgium), a crucial reminder of the strategic and operational parameters set by the international context, and one that was to be repeated in 1741–3.

Conversely, the Franco-Sardinian invasion of Lombardy in the winter of 1733–4 was extremely successful. Good weather allowed the French to cross the Alps, and most of Austrian-ruled Lombardy was rapidly overrun. Austrian attempts to reverse their losses were blocked in 1734 at the battles of Parma (29 June) and Guastalla (19 September). Earlier that year, the Spanish victory at Bitonto (25 June) drove the Austrians from southern Italy, a defeat they were never to reverse. In the battle, the Austrian cavalry was decimated and the infantry were obliged to surrender. The Spaniards followed up by overrunning Naples and successfully invading Sicily. The Russian move into Poland in 1733–4 was another instance of a rapid advance. The Russian captures of Warsaw (1733) and Danzig (Gdansk, 1734) helped lead to the defeat of Stanislaw Leszczyński, the French claimant to the Polish throne, and the triumph of his rival Augustus III, the Elector of Saxony. The attempt by a small French expeditionary force to relieve Danzig was defeated. The Russians went on to move troops into Germany in 1735, but distance and their slow march prevented them from reaching the zone of hostilities before peace was agreed. Earlier, in 1730, the Emperor Charles VI had requested the dispatch of 30,000 Russian troops, and, though they were not sent, awareness of the issue led foreign commentators to follow Russian court politics carefully.

In terms of generalship, there was little new in the war. Marshal Villars led the French invasion of Lombardy in 1733, while, at Philippsburg, the also-elderly Prince Eugene confronted Marshal Berwick (another key figure from the War of the Spanish Succession), who was decapitated by a cannon ball. The poor performance of the Austrians in part reflected Eugene's failure to use the peace years of the 1720s to maintain military effectiveness, although Austria's chronic financial situation was more to blame than any personal failings. To fight the war, the army was increased to 202,598 strong in 1734. The French, in contrast, benefited from their renewed attention to training from the late 1720s, and their operations indicated the strengths of the military machine inherited from Louis XIV, although a failure to sustain the latter's heavy expenditure on the artillery led to the decline of what had been the best artillery in Europe. British and Dutch neutrality ensured that this was very much a land war, although the British staged a major naval mobilisation in support of their unsuccessful attempt to arrange a diplomatic settlement to the conflict. There were no battles at sea.

The war also showed the difficulties of alliance politics and how these were related to strategy. Underlying the tension over strategy, with France opposed to Spain's plan to conquer Naples and Sicily and leave the defence of northern Italy to the French, was an unsuccessful Spanish determination to hold the diplomatic initiative. The peace left Poland to Augustus. The reversion of Lorraine, which had been overrun by the French in 1733 went to the French throne, while Naples and Sicily went to a son of Philip V of Spain.

Eastern Europe, 1735–43

Despite the French success in Italy, it was the continued strength of the Russians which was most notable. In marked contrast to Persia under Nadir

Shah, Peter the Great's achievements were largely maintained by his successors. They did not sustain his drive to make Russia a great naval power, but the commitment to the army was continued and there was also continuity in leadership with Prince Menshikov, General Münnich and Field Marshal Lacy. These generals retained the essentials of the Petrine system and the success they delivered between 1733 and 1742 consolidated Russia's dominance of eastern Europe. Furthermore, at the expense of the Turks, they were more effective than Peter, while Sweden was defeated more rapidly.

Russo-Turkish differences over the Caucasus led to conflict. After an unsuccessful attempt on Azov in 1735, the Russians declared war the following year, seized Azov (after the main powder magazine blew up), and invaded the Crimea, the centre of the power of the Crimean Tatars who were key Turkish allies. Münnich's forces stormed the earthworks which barred the isthmus of Perekop at the entrance to the Crimea; after a bombardment, Münnich ordered a night attack in columns against the western section of the lines. The troops climbed the wall and gained control with scanty losses. However, the Tatars avoided battle in the Crimea, thus depriving the Russians of the battle they sought where they could use their firepower, and the Russians, debilitated by disease and heat, retreated.

Though exhausted by the War of the Polish Succession, the Austrians feared that if they remained neutral they would lose the support of their sole major ally, Russia, and joined in in 1737.[8] Russian ambitions meanwhile expanded, with talk of capturing Constantinople. Münnich took the fortress of Ochakov on the estuary of the Bug in 1737: his advance was supported by supplies brought by boat down the River Dnieper and thence by 28,000 carts. The initial assault failed, but the fortress was taken, again after its powder magazine exploded. The survivors were massacred in the storming of the fortress. Disease (which killed tens of thousands) and logistical problems, however, thwarted hopes of crossing the Dniester and invading the Balkans, and indeed a major outbreak of plague forced the abandonment of Ochakov in 1738. The Russians also encountered considerable problems from the light cavalry of the Tatars: in 1736, when they invaded the Crimea, they suffered from a scorched-earth policy in which crops were burned and wells poisoned, while in 1737 the Tatars burned all the grass between the Bug and the Dniester, hindering Münnich's operations after he took Ochakov.

In 1739, the Russians were more successful. Marching across Polish territory, and thus avoiding the ravaged lands near the Black Sea, Münnich crossed the Dniester well upstream, drove the Turkish army from its camp at Stavuchanakh and captured the major fortress of Khotin and the Moldavian capital of Jassy. The Moldavian nobility pledged loyalty to the Empress Anna. However, like so many bold military schemes of the period, this collapsed not through military difficulties but because of the breakdown of the supporting diplomatic coalition. In 1737, Austrian forces had advanced into Serbia, but they were then driven back, although Turkish attempts to subvert the Habsburg position in Transylvania were unsuccessful. The Turks, nevertheless, ravaged Habsburg Wallachia and Serbia in early 1738, and in June they besieged New Orsova. The Austrians set out to relieve the fortress, defeating the Turks nearby at Cornea (4 July). The Turks then lifted the siege, but, in face of a second Turkish army, and despite another victory near Mehadia, Count Königsegg retreated, abandoning both Mehadia and New Orsova. The Austrian army was hit by plague, divided command and supply problems, and, in general, had declined since the last Austro-Turkish war (1716–18), not only in leadership and discipline, but also in organisation. In particular, there was a use of the standard Western system of linear formations, which was particularly vulnerable to the Turks, whose strength in cavalry enabled them to overcome the Austrian cavalry and then attack the infantry on their flanks. In 1739, at Grocka (Kroszka), advancing Austrian troops under Count Wallis suffered heavy casualties when forcing their way through a defile in the face of the Turkish army. Although the Austrians won control of the battlefield, Wallis responded cautiously by withdrawing. Taking advantage, the Turks besieged Belgrade. The local Austrian commanders, their confidence gone, surrendered, and Austria ceded Belgrade, northern Serbia and Little Wallachia. Abandoned, the Russians felt that they could not fight on by themselves, not least because Sweden, egged on by France, was becoming hostile. Russia did not want to fight a war on two fronts.

Russian prowess was demonstrated more obviously when Sweden, encouraged by France, attacked in 1741. A rapid Russian advance under Lacy led to the successful storming of Willmanstrand in Swedish-ruled Finland. The following year, Lacy attacked again, supported by galleys. The Swedish army was outflanked by a march along a forest route, surrounded in Helsingfors, and obliged to surrender. Fearing amphibious attack on the Swedish mainland, the Swedes yielded to Russian terms in 1743. Finland was returned by the Russians, with the exception of Karelia, and this created a stronger defensive shield around St Petersburg.

The War of the Austrian Succession, 1740–8

Russian successes, however, were overshadowed by those of Prussia. The death of the Emperor Charles VI, ruler of Austria, in 1740 led to a contested succession as the claims for an undivided succession by his elder daughter, Maria Theresa, enshrined in the Pragmatic Sanction, were challenged, first by Frederick the Great, the newly acceded King of Prussia, and later by other rulers as well. Frederick's invasion of the wealthy Habsburg province of Silesia (now in south-western Poland) on 16 December 1740, where his dynasty had territorial claims, was not intended as the opening move of a major European war, a step that would precipitate attempts to enforce claims by other rulers. Frederick hoped that Maria Theresa would respond to a successful attack by agreeing to buy him off, and in many respects his invasion was that of an opportunist. The conquest of Silesia proved relatively easy and substantially increased Prussia's population and resources, as well as providing a crucial geopolitical gain, both in terms of helping protect Prussia from attack and in giving an advantage over Saxony whose elector was also King of Poland. Frederick the Great benefited from inheriting a large and well-trained army from his father, Frederick William I. The unexpected nature of the Prussian advance also helped. At Mollwitz (10 April 1741), the Prussians outnumbered the Austrians by 21,600 to 19,000, and even more in infantry, by 16,800 to 10,000. On the Prussian right, the superior Austrian cavalry shattered their Prussian opponents, and Frederick fled the battlefield to avoid capture. However, his infantry fought well, and were more effective than the Austrians, many of whom were raw recruits. The well-trained Prussians, operating in parade-ground fashion, prevailed over their slower-firing opponents. Prussian casualties were greater than those of the Austrians (4,800 to 4,500), but the Austrian withdrawal led the battle to be seen as a Prussian victory, the first for the largely untested Prussian army.

Maria Theresa refused to cede Silesia, and in June 1741 Frederick signed a treaty with France that committed the latter to military steps. Frederick and France then encouraged other powers to join in, and Charles Albert of Bavaria and Augustus of Saxony did so, as did Spain, each advancing territorial claims, while Charles Albert also pursued the imperial throne. Furthermore, the simultaneous advance of different French forces was an impressive display of coordinated military power. On 21 October 1741, French and Bavarian troops camped at St Polten and Vienna prepared for siege. However, Charles Albert distrusted Augustus and Frederick, fearing that they would seize Bohemia. Instead of advancing on Vienna, he attacked Prague which fell on 26 November to a night-time storming by Bavarian, French and Saxon forces: there was no time for a regular siege and the siege artillery was delayed because of a lack of horses.

As a reminder, however, not of the indecisiveness of campaigning, but of the difficulty of multifront campaigning, this did not mean the collapse of Austria. Instead, Austrian resilience and the possibilities of winter campaigning were demonstrated, as the Austrians recaptured Linz in January 1742, and seized Munich the following month. In turn, Prussian forces invaded Bohemia and defeated Maria Theresa's brother-in-law, Prince Charles of Lorraine, at Chotusitz (17 May): the Austrian cavalry again defeated its Prussian counterpart, but the disciplined Prussian infantry forced the Austrians back. Under pressure from her ally Britain, and keen to lessen the number of her opponents, Maria Theresa bought peace by ceding most of Silesia. Rather than treating this, or the Peace of Dresden of 1745, as ready proof of Prussian military superiority over Austria, it is important to note the extent to which Austria (unlike Prussia) was also heavily engaged against other opponents. The Austrians pressed on to besiege Prague, which the French abandoned in December 1742 in a daring winter retreat that, however, in its exposure of the troops to harsh winter conditions, prefigured that of Napoleon from Moscow in 1812.

British troops had been moved to the Austrian Netherlands in 1742. In 1743, they entered Germany, and, as a clear sign of the role of the monarch as war leader, George II led an Anglo-German army that defeated the French at Dettingen (27 June), a victory gained by the British infantry, whose fire discipline was greater than that of the French. George, however, was unable subsequently to make a major impact on France's well-fortified eastern frontier. The French did not need to use the extensive double crownworks added to the fortifications at Metz and Thionville by Louis de Cormontaigne, who had become Chief Engineer at Metz in 1733. In 1744, Charles of Lorraine was able to cross the Rhine and invade Alsace with Austrian forces, but his efforts were thwarted by a build-up of French strength and by the impact of Frederick's invasion of Bohemia in August 1744.

Frederick thus began the Second Silesian War because of his concern about Austrian strength and intentions, and this was an instance of the extent to which worry about the retention of Silesia was to be a major burden for Prussia, leading Frederick into war. Crossing Saxony, he advanced on Prague with very little resistance and captured it after a short siege. Frederick failed, however, to exploit his success. As after the Franco-Bavarian advance in 1741, there was no decisive battle. The Austrian position near Beneschau was too strong for an attack and meanwhile Frederick's army was being harried by light forces which attacked supply lines and foragers. The Austrian introduction of these light troops from their Balkan frontier was a new development that caused much comment. At the end of the campaign, Frederick retreated, having suffered heavy losses. There had been no repetition of the crisis of Austrian power in 1741.

In 1745, the Austrians, with Saxon support, in turn took the offensive. At Hohenfriedberg (5 June), however, a Prussian riposte, centring on an infantry advance supported by the now more aggressive Prussian cavalry, was successful. The attack has been described as oblique, but can also be seen as flanking. Frontal infantry attacks later that year brought Prussia victory at Soor (30 September) and Kesseldorf (15 December). These victories appeared to vindicate the commitment to cold steel that had led Frederick in 1741 to order his infantry to have their bayonets permanently fixed when they were on duty. Prussian successes enabled Frederick to capture Dresden and to retain Silesia in the subsequent Peace of Dresden.

Meanwhile, full-scale war between France and Britain had broken out in 1744. The French were more successful in the Austrian Netherlands than in their storm-dispersed plan to invade Britain. Furnes, Menin and Ypres were captured: the Dutch-garrisoned fortresses were too weak to resist determined attack. In 1745, the British under William, Duke of Cumberland, sought to relieve besieged Tournai: as so often, the attempt to relieve a besieged fortress led to a battle. On 11 May, his infantry assailed the hastily prepared French position at Fontenoy, demonstrating anew their determination and fire discipline, but Fontenoy also revealed what Malplaquet had shown in 1709 and the Prussians were to discover in 1757: the strength of a defensive force relying on firepower and supported by a strong reserve. The earlier failure to capture the French redoubts on the flanks led to the eventual failure of Cumberland's attack. The French infantry was not held down by flank attacks, as it would have been on a Marlborough battlefield, and it redeployed to attack the flanks of Cumberland's column. Cumberland was forced to retreat with far heavier casualties as a result of his brave, but unimaginative and poorly reconnoitred, frontal attack. Fontenoy was rapidly followed by the fall to the French of Tournai, Ghent, Oudenaarde, Bruges, Dendermonde, Ostend and Neiuport. Bereft of the support of a field army, it was difficult for garrisons to prevail against a determined siege.

In a different context, the value of cold steel was to be vindicated within four months of Fontenoy with the victory of the Jacobite Scottish Highlanders over British regulars at Prestonpans near Edinburgh (21 September), a key victory for Celtic warfare. The regular forces only fired one round before the Highlanders, with their broadswords, were upon them. Most of the casualties occurred during the retreat, for infantry formations that lost their order in retreat were particularly vulnerable. The Jacobites subsequently advanced on London, but, discouraged by a lack of support and misled by false information provided by a British spy, turned back at Derby and, having been pursued by the regulars, were finally defeated at Culloden (16 April 1746). There Cumberland's artillery and infantry so thinned the numbers of the advancing Highlanders that those who reached the royal troops were driven back by bayonets. The outnumbered Jacobites were then crushed, and the battle was followed by a harsh harrying of Jacobite areas which was designed to ensure that there was no subsequent rising.

After Culloden, the remains of the Jacobite army assembled at Ruthven, but then dispersed. They would not have been able to mount an effective resistance as they lacked supplies. Guerrilla resistance was more of an option, but the Highlanders would have been penned in between Cumberland and hostile clans, and, without the presence of their commander, Charles Edward Stuart, the resistance now lacked focus. The way was clear for Cumberland's repression.[9]

The operational and tactical mobility of the Jacobite army might suggest a contrast between the more static warfare of slow-moving regular forces and the speed of irregulars; but this would be misleading. One of the most dramatic developments of 1745 was the march of a French army along the Genoese Riviera, its crossing of the Ligurian Alps, and the Franco-Spanish defeat of Charles Emmanuel III of Sardinia at Bassignano (27 September). However, the kaleidoscopic nature of eighteenth-century warfare was such that, the following spring, the Sardinians recovered their losses, and Austro-Sardinian forces defeated the Bourbon army at Piacenza (16 June). This serves as a reminder of the high-tempo nature of campaigning, as well as the extent to which operations could produce a verdict. The Sardinian (essentially Piedmontese) army, which had been 8,700 strong in 1690 and over 24,000 in 1730, had increased to 48,000 in 1748.

The campaign of 1746 also delivered most of the Austrian Netherlands to France. Brussels fell to Saxe on 20 February after a surprise advance. Trenches were opened before Antwerp on 24 May, the garrison surrendering after a week. Mons fell on 10 July after a month's siege. Charleroi was stormed on 2 August, after a brief siege beginning on 28–9 July, and the citadel of Namur capitulated on 1 October. In command of a British-Dutch-German army, Charles of Lorraine was defeated by Saxe at Roucoux (11 October), a battle that centred on the hard-fought storming of three entrenched villages by the French infantry.

In 1747, Saxe outmanoeuvred Cumberland when he sought to regain Antwerp, and defeated him at Lawfeldt (2 July). Saxe's protégé, Count Ulrich Lowendahl, led an army into the United Provinces and rapidly overran the fortresses in Dutch Flanders that covered the Scheldt estuary: Sluys, Sas de Gand, Hulst and Axel fell between 1 and 17 May. Lowendahl then turned to attack Bergen-op-Zoom, one of the strongest fortresses in Europe, the fortifications of which had been strengthened with casemented redoubts by Vauban's Dutch rival, Menno van Coehoorn. Lowendahl began the siege in mid-July, but progress was slow, and the French had to resort on 16 September to the desperate expedient of storming the defences. Massacre, rape and pillage followed. Next year, Maastricht fell to siege by Saxe and an army of over 100,000 troops, fresh evidence both of French military dominance in western Europe and of the professionalism of French siegecraft. The large number of troops involved in the battles (200,000 at Roucoux, 215,000 at Lawfeldt), the fluidity of the fighting, and the extent to which each battle was a combination of a number of distinct but related struggles, anticipated aspects of Napoleonic warfare. The frontage at Roucoux was about 10,000 yards.

Post-war developments

The end of the war, with the Peace of Aix-la-Chapelle (1748), left many issues unresolved, especially Austrian anger over the loss of Silesia, and Anglo-French territorial disputes in North America which were referred to commissioners who would not be able to settle them. This encouraged post-war reforms as an attempt to enhance military effectiveness, and, indeed, more generally to strengthen government. This was particularly pronounced in Austria, where there were fears that, in the next war, Prussia would try to seize Bohemia. The financial and administrative reforms carried out by Haugwitz, and related military changes, made Austria a formidable power prepared for war and better able to afford it than in 1741. The military reforms did not seek to copy those of Prussia, though they were designed to confront them. One of the most effective was the reform, standardisation and improvement of the artillery by Prince Joseph Liechtenstein, Director-General of the Artillery from 1744, a process the French failed to match. Austrian improvements in their artillery helped to increase their tactical defensive capability against Prussia in the Seven Years' War. In contrast, the French, who followed the artillery methods of Jean-Florent de Vallière, were to be outgunned by the British at the battle of Minden (1759). There was also a marked improvement in the Austrian infantry. New drill regulations, the first for the entire army, were issued in 1749 by a commission headed by Charles of Lorraine, and a military academy opened, with Marshal Daun as commandant, at Wiener-Neustadt in 1752 (the École Militaire in Paris was founded in 1751). Austrian armies were provided with staff officers, and in 1747 a corps of engineers was created. Seven years later, in their camp at Kolin, the Austrians conducted important peacetime manoeuvres.

These changes were related to a longer-term trend in which the Austrian army became less proprietary and more professional. The system of military entrepreneurship, the role in the manning and command of regiments of individuals who used their own resources to support them, had hampered the army's effectiveness and had limited promotion prospects for good officers. This system was gradually dismantled. In 1722, Eugene had complained about the pressure to give regiments to young and inexperienced princes who ran them badly. Maria Theresa was convinced that the dominance of the army by the higher nobility had harmed it. Instead, she sought to create a military establishment financed by regular taxation and commanded by loyal professionals. This entailed a reduction in the military and financial role of the entrepreneurs. After 1744 their opportunities for profit were restricted, while a definite attempt was made to widen the officer corps, in response to the specialised knowledge which was required increasingly, especially in the artillery, and to the limited interest displayed by the higher nobility in such matters and in education in military academies. The Wiener-Neustadt academy was open to the sons of serving officers, a group which included commoners and minor nobles, while the engineers' academy was opened to pupils of all ranks.

The academies created the basis for a professionalised officer corps, producing a service nobility drawn from the middling and lower nobility. In time, service as an Austrian officer became bureaucratic in its nature, and, accordingly, rewarded with prestige, security for old age and guaranteed employment, not lands and lordship. Promotion within the officer corps, previously largely controlled by regimental commanders, was gradually transferred to the government. The service-nobility nature of the Austrian officer corps was further enhanced by the establishment of a political modus vivendi between the Habsburgs and the Hungarians in place of a relationship long troubled by Hungarian wishes for independence, or at least autonomy, wishes that had led to rebellions, most recently in the 1700s. This shift led to a significant increase in the Hungarian element in the army, with new Hungarian regiments, especially in the infantry.

Maria Theresa was determined to ensure the loyalty of the new professional officer group. She consistently sought to upgrade its self-image and social standing, ordering in 1751 that all officers were to be admitted to court. At first, officers were ennobled individually for satisfactory service, but, in 1757, she decreed that commoners with thirty years of meritorious service were to be raised to the hereditary nobility, ennoblement becoming an automatic result of service. After the victory over Prussia at Kolin in 1757, the Empress founded the Military Order of Maria Theresa, a graduated scheme of decorations for military service awarded to officers regardless of social rank or religion: 1,241 individuals were awarded decorations.[10]

This dynastic/state control over the concept of honour was one attempted over much of Europe, not least because it helped with one of the major problems of the age: persuading the nobility to govern in the interest of the state. But it was particularly appropriate in the case of military officers. It represented an attempt to disseminate an idea of honour and rank arising from service rather than from birth, one that was especially necessary in armies where it was essential to persuade aristocratic officers to take orders from men who were socially, though not militarily, their inferiors. In Austria, this was linked with Maria Theresa's fairly effective suppression of duelling in the army. Increased professionalism thus had an effect on the officers' code of honour. The military career was seen increasingly in Austria, as elsewhere, as a respectable pursuit and profession, and not just as a mercenary trade or a part-time occupation for gentlemen. This could be related to the burgeoning of military literature from the 1720s.

The mid-century also witnessed reforms that increased the effectiveness of the Russian army. While Shuvalov's proposals for a higher military department or school to provide a sound knowledge of the mechanics and principles of war came to nothing, the Military Commission created in 1755 produced new regulations for the infantry, cavalry and Cossacks. The infantry code published in 1755 stipulated Prussian-style tactics. In 1756, the artillery held a number of long exercises which improved speed and accuracy; in 1757, it was reorganised, and in the late 1750s received a series of new pieces. These reforms gave the Russian army greater firepower in the Seven Years' War and the artillery a sound professional basis.

The same emphasis on preparation and change was not matched throughout Europe, although there were examples of other states taking major initiatives to improve their military effectiveness, Charles Emmanuel III of Sardinia carrying out a major reform in 1751. In 1753, a Bavarian official, commenting on the new regulations for the Bavarian army, wrote, 'It is today the hobby of all the major princes.' Some of the minor states followed suit, the republic of Genoa, for example, reorganising the inspection of its fortifications in 1748. This process was also seen at sea. The Marquis of La Ensenada, the key Spanish minister, expanded shipbuilding facilities at Cadiz, Cartagena, Ferrol and Havana after the end of the war, hired British designers and craftsmen, and increased the size of the navy, as well as turning out more stable ships. France also greatly increased its navy, improving the design and construction of their warships and the production of naval ordnance.[11]

Such expansion, however, was extremely expensive, and this contributed to the ever-widening gap between first rank and other powers. Only the former had the capacity to play a major independent role in warfare, and therefore in international relations. The ability of states' financial-administrative systems to bear the increased burden of war was crucial if they were to remain first rate. Thus, after 1748, Frederick the Great was able to establish supply depots throughout Prussia in order to speed future mobilisation. Weaker states suffered from the contributions and other exactions which resulted from the deployments of their more powerful counterparts. For this reason, in 1734, the Elector of Cologne opposed the wintering of Prussian troops in his territories.

Conclusions

At sea, only Britain, France and Spain (in that order) were first-rank powers, and the Dutch could no longer compete. Russia, Sweden and Denmark were locally important in the Baltic, but not of wider significance. Austrian interest in the 1720s in developing Ostend and Trieste as ports had not led to the creation of a navy. When he added East Friesland to his possessions in 1744, Frederick the Great acquired the port of Emden, although the Prussians had garrisoned it since 1682. Emden gave easier access to the Atlantic than the earlier Prussian ports on the Baltic. However, Frederick lacked the resources and interest to develop Prussia as a naval power and had a smaller navy than Frederick William, the Great Elector, had had in the 1680s. A small Prussian flotilla, launched at Stettin in 1759, was destroyed by a larger Swedish squadron that September. In the 1750s, Frederick also sought to compensate for his naval position by licensing privateers, the typical action of a weak naval power. On land, formerly important 'players' in international relations, such as Bavaria, Denmark, Sardinia, Saxony and Sweden, could no longer hope to compete with the first-rankers. Whereas the second-rank powers had played a major role in the War of Austrian Succession, the Seven Years' War (1756–63) was to be an epic struggle for the first-rankers.

7

WARFARE, 1756–74

The late 1750s brought to a close the long-standing series of conflicts between China and the Zunghars of Xinjiang. Victory underlined China's position as the leading land power, but this was a war waged entirely on land and only in East Asia. The contrast with the range and variety of the Anglo-French struggle was striking. Indeed, the naval strength and transoceanic advances of the British at the expense both of the French in North America in 1758–60 and of the French and native powers in India in 1750–65, ensured that Britain became the strongest global power, and underlined the extent to which the Seven Years' War was a world war, an important contribution to the global context of military developments.

Within Europe, there were no struggles that were as striking in their consequences, but that does not mean that it is pertinent to appropriate the warfare of the period as indecisive, and that is the case whether the wars are considered in themselves or as an indicator of a more general situation. The reiterated claim about the indecisive character of pre-French Revolutionary warfare is questionable. Aside from individual battles or campaigns, the diplomacy of the period, a vital source for contemporary assumptions about military capability, provides, with its bold plans for alliances and partitions, little sense of military indecisiveness.

The Seven Years' War in Europe, 1756–63

Yet, such an argument has been made, not only more generally for *ancien régime* warfare, but also, more specifically, for the mid-century. The latter has been linked to developments in weaponry, particularly the spread of artillery. It has been claimed that, during the Seven Years' War (1756–63), the impact of cannon and howitzers turned the artillery into the 'dominating force on the battlefield,' and that this was used to deter attack, making war more indecisive: 'the respect induced by the improved efficiency of artillery fire was to keep the enemy in check ... Trapped ... armies were facing each other, until one or other ran out of supplies.'[1] Artillery could indeed be devastating in defence, but (as with sieges) it could also be a battle-opening tool. Furthermore, the failure to crush Prussia in this war did not prove the indecisive nature of campaigns, or indeed war, but, instead, that Prussian campaigning had the decisive effect of securing its survival, eventually in a synergy with a key political shift in one of its most dangerous opponents, Russia.

Maria Theresa of Austria (r. 1740–80) wanted to regain Silesia, while her ally Elizabeth of Russia (r. 1741–62) saw Frederick the Great as the principal obstacle to her aspiration to dominate eastern Europe. In March 1756, the Russians produced a plan for war, but Austria persuaded Elizabeth to delay the attack until 1757. Well aware of Austro-Russian military preparations, Frederick decided to ignore the advice of his new ally Britain to restrict himself to defensive moves and, instead, to launch a pre-emptive strike with his well-prepared army. In order to deny a base to the gathering coalition against him, to gain resources, and to obtain more room for manoeuvre, Frederick invaded Austria's ally Saxony on 29 August 1756. Its capital, Dresden, fell on 9 September and the Saxon army capitulated at Pirna in October, and was then forcibly enrolled in Prussian service.

This invasion was a dangerous move. Louis XV of France felt obliged to succour his heir's father-in-law, Augustus III of Saxony, and Frederick found himself in an increasingly desperate situation. In January 1757, Austria and Russia concluded an offensive alliance, and, in May, Austria and France followed, France promising an army of 105,000 and a substantial subsidy to help effect a partition of Prussia. Sweden and most of the German rulers joined the alliance, while Britain, at war with France since 1754 (although it was not declared until 1756), was not in a state to provide much assistance. Frederick was fully conscious of his vulnerable position and, in January 1757, compared himself to Charles XII of Sweden at the beginning of his reign when three neighbouring powers (Russia, Saxony-Poland and Denmark) had plotted his fall.

On 13 September 1756, Frederick invaded Bohemia, but the Austrians put up unexpectedly strong resistance. At Lobositz (1 October), their artillery was particularly effective and, although Frederick won a tactical victory, the Austrian infantry fought better than before. The Prussians withdrew from Bohemia a fortnight later, not the most encouraging start to the war. The summer and autumn of 1757 were a period of particular difficulty, with a Russian invasion of East Prussia and victory there at Gross-Jägersdorf (30 August), a Swedish invasion of Pomerania, the French conquest of Hanover after defeating a British-financed army under William, Duke of Cumberland at Hastenbeck, the raising of the Prussian siege of Prague and the end of the Prussian invasion of Bohemia after the Austrian victory at Kolin (18 June), a successful Austrian raid that captured defenceless Berlin briefly on 16 October and the Austrian capture of most of Silesia, including its capital Breslau on 25 November. Frederick, however, saved the situation at Rossbach (5 November), inflicting greater losses on a French force that outnumbered his, before using the oblique attack to repeat the experience at Leuthen (5 December) at the expense of the Austrians. In both battles, the Prussians had a certain element of surprise: the attacking force emerged unexpectedly from behind hills.

At Rossbach, Frederick, with 21,000 troops, attacked the French (30,200) and the army of the Empire (10,900), which had planned to turn the Prussian left flank. Responding rapidly, Frederick attacked his opponents on the march, screening his move behind a hill. Major-General Seydlitz surprised and defeated the opposing cavalry of the Allied advanced guard, attacking them in front and with a double-flanking movement. The Allied cavalry was pushed back and dissolved into a confused mass. Seydlitz was an unusually impressive cavalry commander, able to keep control over his men. He then rallied the Prussian cavalry. The advancing columns of French infantry were rapidly brought low by salvoes of Prussian musket fire, supported by a battery of eighteen Prussian heavy cannon. Seydlitz's cavalry then attacked the French infantry and was joined by the advancing Prussian infantry, firing as they moved. The French fled in confusion, covered by their light infantry. The Prussians lost fewer than 550 men, their opponents more than 10,000, many of them prisoners. Frederick's ability to grasp and retain the initiative, and the disciplined nature of the Prussians, both infantry and cavalry, were decisive.

Rossbach is usually cited as a paradigm shift, with France passing the military baton to Prussia as the new exemplar of good practice. France, however, was defeated as much by its own poor logistics and appalling generalship as by Prussian superiority. This underlines the difficulty of assessing effectiveness and also the definition of a military system or culture. Logistics and generalship (the French commander Soubise was a patronage appointment) were as much a part of a system/culture as organisation, tactical doctrine and weaponry.

Rossbach secured Frederick's western flank, persuaded George II to order the resumption of fighting the French in Hanover (which was important in saving the Anglo-Prussian alliance) and seriously challenged the prestige of the French army and monarchy. Thereafter, the French were far more cautious about acting against Prussian forces. The movement of British forces into the region in 1758 further strengthened Frederick's flank and led in 1759 to a British victory over the French at Minden. However, the French did better than is generally allowed in the Seven Years' War: Rossbach and Minden unduly eclipse their successes against Anglo-German forces in Westphalia, particularly in 1760 when they captured Kassel.

At Leuthen (5 December 1757), Frederick, with 35,000 troops, advanced to attack the Austrians under Charles of Lorraine with 54,000. Frederick, crucially, took and retained the initiative. Benefiting from the cover of a ridge, the Prussians turned the Austrian left flank, while a feint attack led the Austrians to send their reserves to bolster their right. The Prussians were helped greatly by mobile artillery. Charles wheeled his army, creating a new south-facing front stretching through the village of Leuthen. The second phase of the battle centred on repeated Prussian attacks on this new front, especially on Leuthen, which was finally carried after bitter fighting, but the

Prussian infantry became exposed to the Austrian cavalry. An Austrian cavalry counter-attack under Lucchese was prevented from reaching the open flank of the Prussian infantry by the prompt action of their cavalry, and the battered Austrian infantry finally fled. This was a hard-fought victory by a well-honed army, which reflected Prussian firepower, Frederick's skilled exploitation of the terrain, the fighting quality of the Prussian cavalry and the ability of Prussian commanders to take initiatives. The Prussians lost 6,380 killed and wounded, the Austrians 10,000, and 12,000 prisoners. After their defeat, the Austrians abandoned most of Silesia.[2]

In 1758–9, pressure from Russia was most acute. In January 1758, the Russians captured East Prussia, which they were to hold for the rest of the war. They pressed on to invade Brandenburg, only to be blocked in the Battle of Zorndorf (26 August), in which Frederick lost a third of his force and the Russians 18,000 men. The following year, better artillery helped bring the Russians victory at Paltzig (23 July), while the Russians defeated Frederick at Kunersdorf (12 August), the Prussians losing nearly two-thirds of their force. These defeats helped to fortify Frederick's already strong fear of the Russians, one that he attempted to assuage by disparaging remarks about them, as the presentation of his battles was important to his prestige, but which conditioned his policies for the remainder of his life.

By the end of the war, the Russian army was the most powerful in Europe; its ability to campaign successfully in Germany was displayed in a conflict with Prussia, which had itself overthrown the image of French superiority at Rossbach. The Russian army made operational and tactical progress, including in the use of field fortifications and light troops and the handling of battle formations. The adoption of more flexible means of supply helped to cut the baggage train of the field army, making it less like that of an oriental host, and the daily rate of march increased, although logistics remained makeshift, and this had a serious impact on operations, for example forcing the army to retreat to the Vistula in late 1760.[3]

The value of the Russian contribution was, however, compromised by the divided nature of the alliance, which greatly compounded the Prussian advantage of fighting on interior lines, providing opportunities for the Prussians to defeat, or at least fight, their opponents separately. The Austrians focused on Silesia, and, after Rossbach, the French on Westphalia. The marked reluctance to coordinate operations was shown by the Russian failure to follow up Kunersdorf by concerted action with Austria. In 1760–1, the Austrians, whose effectiveness became more apparent during the war,[4] consolidated their position in Saxony and Silesia, while the Russians seized Berlin temporarily (9–12 October 1760) and overran Pomerania: Kolberg surrendered on 16 December 1761.

The cumulative strain of the war was serious for Prussia, and casualties were very heavy, contributing to the attritional character of the conflict. Political will was all-important. Frederick should have been beaten after Kolin

(1757), and especially after Kunersdorf (1759), but neither defeat knocked Prussia out of the war. After defeats, each state resolved not to give in but to fight on. Kolin only decided that it would be a long war. This reflected both political factors and the difficulty of replacing well-exercised dead soldiers with equally drilled new recruits, who, anyway, lacked combat experience.

Frederick was saved by the death on 5 January 1762 of his most determined enemy, Elizabeth of Russia, and the succession of her nephew as Peter III. Frederick was his hero, and he speedily ordered Russian forces to cease hostilities. On 5 May 1762, a Russo-Prussian peace restored Russian conquests, and Sweden followed on 22 May. Peter's assassination in July, and the succession of his wife, Catherine II (the Great, r. 1762–96), was followed by a cooling in Russo-Prussian relations, but Catherine did not wish to resume the war. Austria was now isolated and driven from Silesia. The battle of Burkersdorf (21 July) broke Daun's will to continue, and Frederick's victory at Freiberg (28 October) gained most of southern Saxony. On 15 February 1763, Austria was obliged to sign peace at Hubertusberg on the basis of a return to pre-war boundaries.

The Seven Years' War overseas

The Peace of Paris provided a very different verdict for Britain's simultaneous conflict with France and Spain. The key advantage was that the British focused on transoceanic operations, where the combination of their naval strength (underlined by victories over the French fleet in European waters in 1759), and experience in amphibious operations, provided them with options and an ability to follow them through plans. Furthermore, prior to 1762 they were opposed to France, not Spain as well, and French colonies lacked the strength, particularly population and agricultural resources, of mainland Spanish colonies, while the latter were also less accessible for the British due to distance and disease.

Conflict had begun in 1754 when a force of Virginia militia under George Washington, dispatched to resist French moves in the Ohio Valley, was forced to surrender at Fort Necessity. The British vigorously responded the following year, and war was declared in 1756. In 1758–60, Britain conquered French Canada, in 1758 the French bases in West Africa, in 1760–1 those in India, and in 1759–62 much of the French West Indies. The year 1762 also brought the capture of Havana and Manila from Spain, an achievement that was to be repeated by the USA in 1898.

These triumphs reflected the ability to combine local and distant resources. Thus, North American militia, Native American allies and Indian sepoys all contributed manpower, while food, forage, water and transport (mules, bullocks, oxen, wagons) were acquired locally. However, it was also necessary to bring troops and munitions, especially cannon, on long, hazardous and unpredictable journeys from Britain. The ability to do so was crucial to British military capability and was enhanced by the establishment of storage points: garrisons from which troops and munitions could be obtained. Both the ability to move forces long distances and the availability of local support improved the confidence of British military planning. No other state in the world could match this capability: none had such a military system, and therefore none could share these goals.

Alongside an emphasis on structural strengths, particularly fiscal resources and naval power, it is also appropriate to note other factors, not least when considering serious British failures. These included the successful French–native ambush of Braddock's larger army at the Battle of the Monongahela on 9 July 1755, Admiral Byng's failure to relieve the garrison of Minorca in 1756, the loss of Forts Oswego, George and William Henry to French advances from Canada in 1756–7, the abortive British plan to capture Louisbourg in 1757, the heavy losses in the poorly managed and unsuccessful British frontal attack on Carillon (Ticonderoga) in 1758, the French victory at the battle of St Foy over the British army outside Quebec in April 1760, and, in India, the French capture of Fort St David in 1758.

British failures serve as a reminder that the French were helped by the difficulties of the British task, not least the complications of amphibious operations, the problems of operating in the interior of North America, the need to allocate limited resources across a number of schemes, logistical issues and the resourcefulness of the leading French commanders: Lally in India and, even more, Montcalm in Canada. Major British successes were frequently only obtained with considerable difficulty. Thus, the capture of Quebec in 1759 followed a frustrating two months in which the natural strength of the position, French fortifications and the skilful character of Montcalm's dispositions had thwarted the attacking British force under James Wolfe.

The relatively small forces involved in transoceanic operations and the close similarity of their weapons and methods of fighting put a great premium on leadership, not least an ability to understand and exploit terrain, as well as morale, unit cohesion and firepower. The British were generally adept at all of these, but so also were the French, and sometimes more so. Montcalm, who made effective use of French troops and native allies, understood how best to operate in the interior of North America and was particularly successful in 1756–7.

The duration of the struggle was significant. Had the war ended in 1758, or even the end of April 1760, then it would not have been anywhere near as successful for the British. The interconnectedness of the war was also as important as the conflicts in particular areas, with the British able to move forces within their imperial system. In 1762, French colonies in the West Indies fell to British troops sent from North America, rather than distant Britain, and whereas the French force that temporarily captured St John's in Newfoundland that year had to come from France, the British force that drove it out came from North America. The crucial interconnectedness of

British power was that naval predominance and success in European waters meant an ability to grasp the initiative further afield. Although blockade was a difficult task (see pp. 160–61), that of Brest, the Atlantic base of the French navy, made it hard for France to send substantial reinforcements to their colonies or to maintain important trade links within them. Thus, the French imperial system was seriously weakened before the British captured the colonies. Maritime links were vital for demographic, economic, organisational and military factors. Furthermore, thanks to British naval strength, French attacks, such as Lally's siege of Madras in 1758–9, or the siege of Quebec in 1760, could be more readily thwarted. British advantages also made it easier to exploit victories, such as the defeat of the French army in India at Wandewash in 1760, which left their surviving bases vulnerable to attack.

The French deployed fewer military resources than the British in their colonies, in part because their war-hit economic and fiscal systems could only support so many troops. In 1758, the British had 24,000 regulars and 22,500 provincial troops in North America, the French about 24,500 including militia. The size of the respective forces was not the sole determinant of success. The Canadian militia was very good and had more experience in wilderness warfare than their New England opponents, the large British armies faced logistical problems, and in the operations near Quebec in 1756–60, and on Martinique in 1760, it was not numbers alone that were at issue. Nevertheless, the exceptionally large resources devoted to the struggle in Canada by the British in 1758–60 stacked the odds against New France. The war left Britain with its gains of Canada, Senegal, Grenada, Tobago, Dominica and St Vincent recognised by France, while she gained Florida from Spain in return for the return of Cuba.

War with non-Europeans

Conflict in North America in the early 1760s, with the Cherokee in North Caroline in 1760–1 and, even more, Pontiac's War in 1763–5, indicated, however, that British proficiency against other Western forces, particularly the French (see p. 97) who, lacking a populous hinterland, were vulnerable to amphibious attack, was no guarantee of success in confronting non-Western forces. This was also exemplified in South Asia. In India, the British were successful in Bengal in the early 1760s, but very much less so against Mysore in the late 1760s, while, in 1761–6, the Dutch, who had been the European colonial power in the coastal areas of Sri Lanka since replacing Portugal in the mid-seventeenth century, faced a difficult war that indicated the limitations of the Western military. Far from the war beginning with an act of Western expansionism, it was launched by Kirti Sri, the ruler of the interior kingdom of Kandy. Exploiting discontent in the militarily weak Dutch coastal possessions, he attacked and overran most of the coast. As elsewhere, however, where Western power was attacked, for example in North America during Pontiac's

War, and in India in the face of Maratha or Mysore attacks, it proved harder for the indigenous forces to capture fortified positions, and Negombo successfully resisted attack in 1761. Furthermore, the Dutch, like other Western colonial powers, benefited from their ability to deploy troops from elsewhere in their empire: many of the Dutch reinforcements came from the East Indies. By the end of 1763, they had regained the coastal regions.

In 1764, however, when the Dutch set out to take the interior, dispatching six columns against the capital, they were as unsuccessful as the much earlier Portuguese expeditions into the interior in 1594, 1630 and 1638. There had been no improvement in Western offensive military capability in the meanwhile, and the usual problems of operating in the tropics, particularly disease, difficult terrain and an absence of maps, were exacerbated by Kandyan resistance. Taking advantage of the jungle terrain, Kandyan sharpshooters harassed the Dutch, inflicting heavy casualties.

Learning from past mistakes, however, was an important characteristic of successful Western operations. In January 1765, the Dutch launched a new campaign, replacing swords and bayonets with less cumbersome machetes, providing a more practical uniform and moving more rapidly. To begin with, the Dutch triumphed, capturing the deserted capital, but the Kandyans refused to engage in battle – always a sensible response to Western firepower. This meant that Dutch energies were dissipated in seeking to control a country rendered intractable by disease and enemy raiders. Peace was made in 1766, with the Dutch accepting that they could not control the interior. Similarly, although Dutch bases in Sri Lanka were captured by amphibious British forces in 1782 and again in 1795–6, on the other hand, in 1803 the British were thwarted, in war with Kandy, by guerrilla attacks, logistical problems, inhospitable terrain and disease. The garrison in Kandy was obliged to surrender in June and was then massacred on its retreat to the coast. In 1815, in contrast, when war resumed, opposition was overcome as a result of concerted operations by independently moving British columns.

Eastern Europe, 1764–74

Military historians generally jump from 1763, the end of the Seven Years' War, to 1775, the outbreak of the American War of Independence. None of the Western European powers was involved in a major conflict during this period. The British used intimidatory gunboat diplomacy in 1764–5 to defend their position successfully in colonial disputes with France in West Africa and the West Indies, over Gambia and the Turks and Caicos Islands respectively, and in 1770–1 Spain and France backed down in a confrontation with Britain over the Falkland Islands, from which, in order to maintain territorial claims, Spain had expelled a British garrison.

It is, however, instructive to consider the warfare of these years, not least in eastern Europe. There the focus is on the Russo-Turkish war of 1768–74, but the variety of conflict in this period was shown in the War of the Confederation of Bar between Russian regulars and Polish patriots from 1768, which culminated in the First Partition of Poland in 1772. Russian attempts to suppress Polish independence of action was assisted by the divisions amongst their opponents in Poland, but the campaigns were far from easy. The problems of controlling a vast territory were exacerbated by the mobility of the Polish light cavalry, while the decentralised nature of Polish politics ensured that it was not possible to win the war by identifying and capturing a small number of targets, although the Poles were, in turn, affected by a serious lack of unity, as well as poor logistics.[5] General Dumouriez, who was sent in 1770 by the French government to create a Polish army that would, with French help, be able to resist Russia, found the Confederates divided, badly armed and disciplined and less numerous than he had hoped.

Operationally, the Russian emphasis was on resources and mobility: Russian success depended on the remorseless deployment of substantial numbers, as in the successful three-month siege of Cracow in 1772, and on their willingness to force rapidly moving engagements by bold attacks. At Landskron in 1771, Russian infantry and cavalry stormed the Polish position; in the ensuing battle, the cavalry put the Polish infantry to flight while the Russian infantry held off the Polish cavalry. At Stalowicz (12 September 1771), a bold, surprise, dawn advance into the village where the forces of Lithuania were based brought the Russians, under Alexander Suvorov, victory. More generally, his successful emphasis on speed indicated that, far from being simply formulaic, as might be suggested by volley training and linear formations, there was a dynamism and flexibility in *ancien régime* European warfare. Moreover, in a significant shift in the distribution of power, the partitioning powers, Russia, Austria and Prussia, acquired major territorial gains from Poland with little resistance. In contrast, none of the partitioning powers had gained territory in the Seven Years' War.

Russian effectiveness was also shown at the expense of the Turks. Concerned about the potential consequences of Russian control of Poland in any future war, and encouraged by France and by the Tatars, the Turks responded to Russian violations of Polish territory by declaring war in October 1768. However, it was to be the Russians who repeatedly took the initiative. In 1769, Prince Alexsander Golitsyn advanced to the upper Dniester and captured Kamenets Podol'sk, Khotin and Jassy. Criticised, however, for a lack of aggressive drive, Golitsyn was replaced by Count Peter Rumyantsev, who had played a major role in the Seven Years' War. Greatly influenced by Frederick the Great, Rumyantsev was a firm believer in the offensive, both operational and tactical. He abandoned traditional linear tactics and, instead, organised his infantry into columns that could advance rapidly and independently, and re-form into hollow divisional squares, while affording mutual support in concerted attacks. In this way, the mobility of Turkish cavalry, which so threatened forces deployed in a linear fashion by turning their flanks, was overcome. The columns relied on firepower to repel Turkish assaults and included mobile light artillery. A major role was also played by bayonet charges: firepower was followed by hand-to-hand fighting. These tactics prefigured those of the forces of Revolutionary France, although the Russians tended to compartmentalise their military experience and to argue that conflict on their southern steppe required distinctive practices to those necessary for victory against Western foes such as Prussia.[6]

Rumyantsev's tactics helped to bring success in battles such as Ryabaya Mogila, Larga and Kagul in 1770, and Kozludji in 1774. Turkish casualties were far greater than those of the Russians: at Larga, 3,000 Turks to fewer than 100 Russians; at Kagul, 20,000 to 1,470. Success was more than a matter of battlefield skill and determination. The Russian army was also increasingly expert in the deployment of their forces. The adoption of more flexible means of supply helped to reduce the cumbersome baggage trains, although logistics remained a serious problem until the development of railways in the nineteenth century, not least because of the, by modern standards, primitive nature of the empire's administrative system. However, improvements permitted better operational planning, including more effective use of riverine and littoral communications. Aware of the logistical difficulties faced by a large army, Rumyantsev grasped the need to take the initiative.

In late 1769, he sent units forward into Wallachia and into Moldavia, where the Russians were supported by the local population. The Russians advanced as far as Bucharest; further east, Azov and Taganrog were also secured. Rumyantsev's army wintered in Polish territory between the Dniester and the Bug. In 1770, while Prince Peter Panin overran Bender on the lower Dniester, Rumyantsev advanced down the valley of the River Pruth, successfully storming the main Turkish positions at the battles of Ryabaya Mogila, Larga and Kagul. After his victories over larger forces on the Pruth, Rumyantsev advanced to the lower Danube, where he breached the Turkish fortress system, rapidly capturing Izmail, Kilia and Braila. Akkerman and Bucharest also fell. The Turkish Grand Vizier, Mehmet Emin Pasha, lacked military competence, had no effective plan and was unable to arrange adequate supplies or pay for his army. The way to Constantinople seemed clear.

Also in 1770, a Russian fleet, which had sailed from the Baltic in 1769, attacked and destroyed the Turks at Cesmé. The Russians were then able to blockade the Dardenelles, although their attempts to capture the islands of Lemnos, Euboea and Rhodes were unsuccessful. Encouraged by Russian promises of assistance, the Greeks in the Morea (Pelopennese) rebelled, but the Russians failed to provide their promised support and it was difficult to coordinate Greek action. The Turks were able to suppress the revolt. Thus, Russian naval power did not have the anticipated strategic effect.

In 1771, the Russians overran the Crimea, installing a pliant khan, but, thereafter, Russian forces were distracted by the First Partition of Poland (1772) and the large-scale Pugachev serf rising within Russia itself (1773–5), and in 1772 attacks were suspended during negotiations. In 1773, the Imperial Council decided that Rumyantsev should attack the main Turkish army south of the Danube, although the general was unhappy about doing so with a small army and with communications threatened by Turkish garrisons, especially Silistria on the Danube, which resisted assault. Accordingly, having crossed the Danube in June, Rumyantsev retreated in the following month.

In 1774, he crossed the Danube anew. The main Turkish army, a much larger force, was routed at Kozludji (9 June): the Russian square advanced and beat off a janissary attack that was supported by Baron Tott's batteries, part of the Turkish borrowing from Western warfare. In the battle, rain spoiled the cartridges in the cloth pockets of the janissaries; the Russians, who used leather pockets were more fortunate. Russian firepower was supported by a cavalry attack that broke the Turkish will to fight; twenty-five of Tott's cannon were captured. As a result of this victory, the main Turkish fortresses were left isolated and vulnerable, a reminder of the extent to which fortresses were most useful if supported by the presence of a field force. The Turks hastily made peace.

By the Treaty of Kutchuk-Kainardji of 1774, the Russians gained territory to the north of the Black Sea, including the coast as far as the Dniester. The war had amply demonstrated Russia's military prowess, and, after the abortive starts in the 1680s and 1710s, really began the 'Eastern Question': the international power politics that focused on the fate of the Turkish empire. Indeed, this issue suggested that some of the traditional diplomatic agenda was redundant.

Corsica, 1768–74

At a very different scale, the French conquest of the Mediterranean island of Corsica in 1768–9 provided another instance of asymmetrical warfare in Europe. French occupation of Corsica, following its purchase from the Republic of Genoa in May 1768, was resisted in part due to exactions by French forces. There was a long tradition of resistance to Genoese rule, and this had led in the 1730s to French intervention on behalf of Genoa. In 1768, French overconfidence and poor planning, combined with Corsican resolve, knowledge of the terrain and fighting qualities, led to defeat for the French at the Battle of Borgo (5–9 October): the attempts to relieve the surrounded French garrison at Borgo, and the latter's to break out, were repulsed, and, on 10 October, the garrison of 530, plus twenty cannon, surrendered. Later in the year, the French forces, suffering constant harassment from the Corsicans under Pasquale Paoli, were driven back.

In early 1769, however, French forces were increased to 24,000, and their already-active programme of road construction was stepped up. At a *consulte* held that March in Casinca, the Corsicans responded by requiring all able-bodied men between the ages of sixteen and sixty to serve in the nation's defence and by voting to defend themselves until death. The French, under the Count of Vaux, planned a coordinated three-pronged attack, which engaged Paoli while threatening his line of retreat. Paoli responded, at the Battle of Ponte-Nuovo on 8 May, by attacking part of Vaux's force, only to be caught in a heavy crossfire from other French units. Paoli's defeat with heavy casualties was followed by his retreat through Corsica and flight abroad on 13 June. Although the cause was fashionable with liberal opinion, Paoli failed to win international support. The British, crucially, were unwilling to risk war with France, and, although concerned, Charles Emmanuel III of Sardinia also would not act. Guerrilla-type resistance continued, and the French responded with devastation, terror and road construction. Opposition was met by burning grain, cutting vines and uprooting olive trees. Those found carrying arms were killed by mobile columns, and, by the spring of 1770, Corsica had been subdued. A fresh rebellion in 1774 was stamped out by the use of collective punishments, including the burning down of villages.[7]

Although the forces involved were relatively small, the Corsican campaign was instructive, but it did not have an important influence on French military thinking. Bourçet advised Vaux, while Napoleon (who, born in 1769, was, far more significantly, a French subject because of the incorporation of Corsica) studied the campaign and was shown the battlefield of Ponte-Nuovo by Paoli. However, French counter-insurgency operations in 1793–1813 owed little to the operations in Corsica. More generally, the campaign, like the successful expulsion of Austrian forces by the Genoese in 1746 and the subsequent resistance in Genoa to fresh occupation by Austria, indicated the strength of determined popular action. The 1769 campaign also showed, however, like that of 1746 in Scotland, that irregular forces could be defeated tactically by superior firepower, especially if they attacked prepared opponents. Furthermore, it indicated the potential of coordinated independent regular forces operating against irregulars.

Conclusions

The Corsican campaigns also demonstrated the strength of major states. France could mount the logistical effort required to deploy a considerable force on an island where provisions were in short supply. She could also sustain defeat and yet return to the attack, proceeding systematically to obtain a planned military outcome. The construction of roads was symptomatic of the entire process. The French army had the engineering skill and manpower to build roads that could serve direct military purposes, the movement of men and, more crucially, artillery and wagon-borne supplies, as well as extend the range of routine authority. Similarly, there had been road-building in Scotland in the 1720s and after the suppression of the 1745 rebellion. This might seem a long way from the Seven Years' War, but the common theme was that of the need to add the sustained application of resources in order to gain results from success in battle.

8

WARFARE, 1775–91

American War of Independence, 1775–83

The American War of Independence tends to overshadow other events between the Seven Years' War and the French revolutionary wars in the standard Western approach. It was not the sole conflict in the period, but is important in the global context, and not only because it led to the birth of the USA, the modern military (and academic history) superpower. The war was the first major revolutionary conflict, the first large-scale transoceanic conflict between a European colonial power and subjects of European descent, and, from 1778, an important episode in the long-standing struggle between Britain and France. In North America, the war was both revolutionary, in that it was one of the first important 'modern' instances of the 'nation-in-arms', and traditional, in that it was essentially fought on terms that would have been familiar to those who had been engaged in recent conflicts in Europe and North America. The American response to battle, adopting the lines of musketeers of European warfare, was scarcely surprising, as numerous Americans had served in the mid-century British wars against the Bourbons, over 10,000 of them as regulars, while many others were familiar with the methods of European armies, especially the British, through reading, observation or discussion.

Infantry dominated the battlefields in the middle colonies, although the potential impact of the trained British musketeers with the bayonets that inspired fear among the Americans was lessened by the ability of the latter to entrench themselves in strong positions, the value of which was amply demonstrated in the heavy casualties among the British attackers at Bunker Hill (17 June 1775) as they sought to overawe their opponents in the first battle of the war. The terrain of much of America was appropriate for such defences. In place of the open farmland of much of the North European Plain, there were narrow valley routes flanked by dense woodland, deep rivers with few crossing points, and, in New England, the omnipresent stone walls that created ready-made defensive positions. British generals such as Howe and Clinton responded with flanking manoeuvres. Heavily encumbered British regular units, manoeuvring and fighting in their accustomed formations, were not only vulnerable in the face of entrenched positions and unsuited to the heavily wooded and hilly terrain of the Canadian frontier, they were also not ideal for the vast expanses of the South. In that relatively sparsely populated region, supplies were harder to obtain and, aside from the ports, there were fewer places that it was crucial to hold, and therefore less opportunity for positional warfare, which the British sought to force on the Americans. The hot and humid climate of the South also posed problems for the troops.

Alongside that of geographical scale, the political dimension helped ensure that the American war and western European *ancien régime* warfare seemed sufficiently different that the former appeared to have little to teach for the latter. The difference in force–space ratios meant major contrasts in staff work, logistics and operational concepts, while strategy was fundamentally affected by the politics of widespread commitment to independence on the part of the Patriots. This was also true operationally, in the case of the major contribution made by Patriot militia, not least in partisan warfare.[1] This undermined the southern strategy adopted by the British in 1780–1, as it proved impossible to consolidate the British position in the Carolinas.

American generals often used skirmish lines, night marches and hit-and-run attacks, but George Washington not only commanded, in the Continental Army, a force that matched many of the socio-professional assumptions of the British army,[2] but also chose to fight essentially in a manner to which the British were accustomed; they were not obliged to rethink totally their way of fighting on the battleground. This was in marked contrast to the unfamiliar logistical and political problems that were faced. Translated to America, European tactics and fighting quality brought success, strikingly so in 1776, and at the Battle of Camden (16 August 1780), where the British fired accurately as they advanced. British regulars were also better at night fighting than the Americans. The Hessian troops hired by the British were also flexible in their tactics, as at the Battle of Long Island in 1776, where they advanced first in skirmishing order.

The British fought to win, but generally cautious generalship and the absence of cavalry to exploit victory, made it difficult to translate success in the field into overwhelming American losses, and these only occurred when the revolutionaries held a position whence there was no route for retreat, as at Fort Washington in 1776 and Charleston in 1780. As the revolutionaries became more accustomed to battlefield conditions, their ability to repel British attacks, as at Saratoga (1776), or to inflict serious casualties increased, as was shown by the heavy British casualties in their victory at Guilford Court House (1781).

Without reliable popular support in the thirteen colonies that rebelled, the British were obliged both to obtain the bulk of their supplies from Britain and to employ much of their army in garrison duty, an obligation made more necessary by the need to protect supply bases and the crucial trans-shipment points. Thus, only a part of the army was available for operations, while the seizure of new positions, especially New York (1776) and Newport (1776), forced the British to deploy still more of the troops as garrison units. This helps account for the British stress on a decisive battle, because it was only by destroying the American field armies that troops could be freed from garrison duty in order to extend the range of British control. American skill and determination in avoiding such a defeat was therefore crucial.

Initially the war had focused on New England with the British forces concentrated on Boston, and British authority elsewhere in the thirteen colonies overthrown: owing to the concentration of troops in Boston, governors elsewhere were provided with insufficient military support. Although initially successful, an American invasion of Canada failed to capture Quebec, which resisted both siege and storming, but, by the end of the March 1776, after the evacuation of Boston, in response to the threat to the harbour posed by new American artillery positions, there were no British troops in the thirteen colonies. The war seemed completely won by the American Patriots.

In 1776, however, the empire struck back, opening the second stage of the war. The melting of the ice on the St Lawrence enabled a fleet to relieve Quebec, the Americans were subsequently defeated at Trois Rivières (8 June), and they were driven from Canada. The main British army landed at Staten Island on 3 July, defeated the Americans on Long Island on 27 August and captured New York. British troops then advanced across New Jersey, and Congress fled from Philadelphia to Baltimore. British success in the war seemed likely.

A surprise and bold counter-attack on the Hessian position at Trenton (26 December), which involved a tricky crossing of the River Delaware, however, enabled George Washington to regain the initiative and helped stabilise the revolution. During the subsequent war of attrition in New Jersey in January–May 1777, the frequent and devastating attacks by American forces on British outposts and foraging parties demonstrated their capability. Washington's increasing employment of light infantry enhanced American strength, as did the British delay in mobilising the loyalists.

The shift in advantage in New Jersey led the British to mount their advance on Philadelphia in 1777 by the slow sea route rather than overland from New Jersey. This led to a serious delay, and it was not until 11 September that the American force protecting the capital was defeated at Brandywine. Philadelphia fell, but the British advance there was unable to coordinate with the advance south from Canada along the Lake Champlain-Hudson corridor. General Burgoyne's army was rashly exposed to larger American forces and unable to break through their positions at Bemis Heights. Burgoyne surrendered at Saratoga on 17 October, ending any serious prospect of cutting off New England from the rest of America. Furthermore, if the American success in mounting a riposte to the fall of Philadelphia did not lead to victory at Germantown (4 October 1777), it showed that the American forces in the middle colonies were far from defeated.

From 1778, France, which earlier had provided crucial military supplies to the revolutionaries, came formally into the war, in an attempt to reverse the hegemony Britain had established in the Seven Years' War. French intervention added to the range of urgent British military commitments elsewhere in the world. It also threatened British naval control of North American waters and thereby challenged the application of these resources and the articulation of the British imperial system. This was to be followed in 1780 by the dispatch to North America of an expeditionary force under Count Jean de Rochambeau, a veteran of French campaigns in Germany in the War of the Austrian Succession and the Seven Years' War. This ensured that the French threat was not only a naval one. Already, in 1778, French entry had led the British to abandon Philadelphia.

In 1779, France's ally, Spain (which had provided the Americans with gunpowder and bullion via Havana) joined her, and in 1780 the Dutch. The result was a major challenge to British power that both helped distract attention and resources from the war in North America and contributed directly to the British defeat there at Yorktown in 1781: a larger French fleet at the mouth of the Chesapeake blocked the attempt to relieve the army besieged there and it was forced to surrender. This led directly to a crisis of confidence among British parliamentarians, the fall of the government and peace negotiations. Elsewhere, despite losses in the West Indies, West Africa and the Mediterranean and on the Gulf Coast of North America, most of the British empire survived attack, including Gibraltar and Jamaica which had been Spanish goals.

British success in retaining most of the empire in part reflected the problems its opponents faced with coalition warfare, and, as such, mirrored Prussia's survival in the Seven Years' War. Such warfare posed particular problems for command and control. The American War of Independence also reflected the extent to which much Western warfare did not take place in Europe, although exceptionally so for this conflict. The war moreover indicated the narrow margin of success and failure, as can be seen by contrasting the results of Yorktown (1781) and the Saintes (1782). British victory over the French fleet at the latter can be seen as an achievement for a tried naval system (see p. 150), but British failure at Yorktown, and at the hands of the Marathas at Wadgaon (1779) (see p. 17), indicated that the success of the British military system depended on contingencies. So also did the contrast between the success of French military intervention in North America and its failure in India, where the dispatch of troops and warships did not help Mysore overthrow British power in southern India. Had Bussy reached India without delay, with healthy troops and with all his artillery, he would have been more likely to achieve his objectives, not least because Suffren was a better naval commander than de Grasse, his counterpart off North America. However, Bussy had far fewer troops than the 10,000 men he had requested, in part because of the ships which were prevented from sailing or which were intercepted by the British navy. Losses from scurvy also hit French numbers.

Land–sea coordination was a matter of both planning and contingencies. British positions, such as Quebec in 1776 and Gibraltar in 1779, could be relieved by sea, but those that were not relieved, such as Pensacola and Yorktown in 1781, and Minorca in 1782, were lost. Conversely, British initiatives, such as the plan to capture Cape Town from the Dutch in 1781, were thwarted by the arrival of French warships. The war also indicated Spain's continued military effectiveness. The Spaniards captured Baton Rouge, Manchac and Natchez in 1779, and Pensacola in 1781: in the last, a Spanish grenade ignited a gunpowder magazine.[3] The world of Spanish commitments, which extended to Algiers in 1775 and the English Channel in 1779 (both unsuccessful expeditions), was very different to that of Prussia, another reminder of the diversity of the period. The war ended with the Peace of Versailles in 1783. France received Tobago and Senegal, while Spain obtained Minorca and Florida. Yet the major British territorial loss was to the Americans, not the French, and Britain remained the leading European power in India.

European military developments

Alongside an emphasis in this book on chronological segments, it is worth pausing to consider longer-term trends, particularly change within a chronological period not generally seen in those terms, to whit the last stage of the pre-revolutionary *ancien régime*. While rulers and ministers sought to cope with the consequences of the mid-century wars and engaged in highly competitive international relations, commanders sought to gain a capability advantage by responding to what they saw as developments. The development of a reform-minded intellectual climate within armies, albeit not one which was formal in any modern sense, was one of the hallmarks of the period. The Guiberts, Lloyds and Schaumburg-Lippes of the period prepared the way for Clausewitz. Furthermore, commentators such as Henry Lloyd (c.1729–83), who served in the French, Austrian, Brunswick and Russian armies, sought to offer a public analysis of developments. The Seven Years' War attracted particular attention. Lloyd wrote the *History of the Late War in Germany* (1766–90), and John Wilhelm von Archenholz and Georg Friedrich von Tempelhoff also wrote histories of the conflict.[4] The discussion of military theory and practice in the public sphere, a key aspect of what might be termed the military enlightenment, underlines the difficulty of identifying national military cultures in war-making, as opposed to the distinctive problems that led to particular strategic cultures. This was the case for example with 'small war', which was discussed in French works, such as De la Croix's *Traité de la petite guerre* (1752), Lancelot Turpin de Crissé's *Essai sur l'art de la guerre* (1754) and Thomas de Grandmaison's *La Petite Guerre* (1756).[5]

The prestige of Prussian methods was particularly important after the Seven Years' War. Those interested in military matters attended Prussian manoeuvres and studied the campaigns of Frederick the Great. Prussian military regulations were translated and Prussian drill adopted, while those who had served under Frederick were able to gain posts elsewhere, Catherine II of Russia recruiting her relative Count Frederick of Anhalt in 1783. Charles III of Spain used Prussia as a model for both infantry and cavalry tactics, while, in December 1777, Nathanael Greene cited Frederick the Great, 'the greatest general of the age', when attempting to dissuade George Washington from attacking the British forces that had recently seized Philadelphia.[6] From 1764 to 1785, the British kept yearly summaries of the annual Prussian manoeuvres, and the British drill regulations of 1786 were based on a manual by the Prussian Inspector General. Three years earlier, Louis-Alexandre Berthier, later Napoleon's chief of staff and minister of war, was much impressed by the Silesian manoeuvres and applauded Prussian precision. In 1792, Charles Jenkinson commented on a Prussian army review, 'the celerity and precision with which all their movements are performed are inconceivable to those who have not seen them. Every operation they go through is mechanical.' The French envoy in London, in contrast, referred to 'new hordes of disciplined slaves.'[7]

Western commentators remained most impressed by the Prussian army and largely ignorant of Russian effectiveness, as well as of new French ideas. Under the pressure of failure in the Seven Years' War, it was in fact in France that much rethinking took place, though, in addition in 1768, Frederick the Great had considered new, more flexible tactical ideas, in particular an advance in open order.[8] French reforms encompassed tactics, operational ethos, organisation, administration and the individual arms. They began during the Seven Years' War and became more intense thereafter. In 1758, Belle-Isle, the new war minister, regulated military pay and demanded that generals act in a more martial fashion: encamping with their troops and wearing proper uniforms. In 1762, Choiseul sought to reform military administration in order to remove the need to support the costs of units by relying on the credit of their officers. The number of units was fixed, recruiting was transferred to royal agents, and the soldiers took an oath of loyalty directly to the king.

Veterans of the Seven Years' War played a major role in post-war reforms. Jean de Gribeauval, who had served in the Austrian artillery under Liechtenstein during the war, and who became Inspector General of the French artillery in 1776, standardised the artillery, with eight-gun batteries and 4-, 8-, 12-pound cannon and 6-inch howitzers. Mobility was increased by stronger, larger wheels, thanks to shorter barrels and lighter-weight cannon and better casting methods. Accuracy was improved by the introduction of elevating screws and graduated rear sights, the issue of gunnery tables and the introduction of inclination markers. The rate of fire rose through the introduction of pre-packaged rounds. Horses were harnessed in pairs, instead of in tandem.

The theory of war advanced to take note, although within a context of controversy in which change was resisted by conservatives, a common feature of the world of French military development in this period. In his *De l'usage de l'artillerie nouvelle dans la guerre de campagne* (1778), Chevalier Jean du Teil argued that the artillery should begin battles and be massed for effect. Napoleon's thoughts on the use of massed artillery were drawn from du Teil, and he studied in 1788 at the artillery school at Auxonne, which was commanded by du Teil's brother. Also a supporter of such massing, Jacques, Count Guibert, who had fought at Rossbach and Minden, advocated, in his *Essai général de tactique* (1772), living off the land in order to increase the speed of operations and the establishment of a patriotic citizen army. Napoleon was to praise Guibert's writings which, with their stress on movement and enveloping manoeuvres, and their criticism of reliance on fortifications, prefigured his generalship. Under the Count of Saint Germain, Minister of War 1775–7, Guibert reformed military practice. The number of officers was cut and the venality of military offices suppressed. In 1787–9, Guibert served as Director of the Higher War Council. There was also pressure for a less fixed sequence of moves in manual drill and, instead, for a drill that enabled troops to be more flexible, which was seen as a basis for enhanced effectiveness, a contrast to the more rigid drill earlier in the century.[9]

Some discussion was long-standing, particularly that of tactical questions, specifically over the rival merits of line and column formations. In contrast to the customary emphasis on firepower and linear tactics, Jean-Charles, Chevalier de Folard and François-Jean, Baron de Mesnil-Durand emphasised the shock and weight of a force attacking in columns: *ordre profound*, not *ordre mince*. Manoeuvres in 1778 designed to test the rival systems failed to settle the controversy, but the new tactical manual issued in 1791 incorporated both. There was also interest in the development of the division, a unit composed of elements of all arms and therefore able to operate independently. Such a unit could serve effectively, both as a detached force and as part of a coordinated army operating in accordance with a strategic plan. The divisional system evolved from 1759, initially as temporary units, and in 1788 army administration was rearranged along divisional lines: twenty-one combat divisions were created, a development from the concept, introduced by Saint Germain in 1776, of the organisation of France into military zones, known as divisions, with administrative functions, such as recruitment, as well as permanent garrisons. The system of combat divisions, which became a standard wartime procedure for the French in the 1790s, gave generals the potential to control much larger armies than the 60,000–70,000 troops that had been considered the maximum effective force in mid-century. Louis-Alexandre Berthier acted as chief of staff at the camp at Saint Omer in 1788, the year in which the cavalry was reorganised.[10]

At the same time, the general context in the West was of military forces which were more powerful than those of 150 years earlier. In particular, recruitment practices, not least conscription systems, were an important aspect of a long-term rise in state military power and effectiveness, an aspect of the extent to which the need to mobilise domestic resources for war continued to be a key determinant of political and government developments. New recruitment systems were mediated by aristocratic officers, central authority being constrained by the reality of continued aristocratic hegemony, but the systems reflected an enhanced control on the part of government, and there was no longer a figure equivalent to Albrecht von Wallenstein – the independent entrepreneur who raised and commanded armies for the Emperor Ferdinand II in the 1620s and 1630s, who was finally assassinated when his loyalty became (correctly) suspect. Independent military entrepreneurship no longer undermined governmental political and operational control of the military, and this enhanced the ability to think and act effectively in strategic and operational terms. The increase in discipline, planning and organisational regularity and predictability made it less difficult to implement plans. These characteristics were seen not only with the major powers, but also with their lesser counterparts. Although their military tasking were very different to that of the major powers, and in some states, for example those in Italy, especially Tuscany, there was a significant reduction in expenditure and army size, they also sought to emulate the reformist themes seen elsewhere. This was particularly seen with tactical innovations, the improvement of technical branches, especially the artillery, and an interest in military education.

The greater effectiveness of military forces can be variously gauged, and there are counter-examples to most generalisations, but it was certainly more marked by the late eighteenth century than it had been in the mid-seventeenth. For example, Austria's military culture became characterised more by regulation than by improvisation. The emphases of the reform movement of the inter-war period of 1748–56 were maintained: the system of military entrepreneurship centred on colonel-proprietors was qualified by a greater stress on government control and the growth of professionalism. A new transport corps was created in 1771, and the supply system was centralised.[11]

Enhanced organisational capability was not simply a matter of financing, supplying or moving armies and navies, but also improved the organisational and operational effectiveness of individual units. Western militaries moved most towards a large-scale rationalisation of such units: they were to have uniform sizes, armaments, clothing, command strategies, etc. Such developments made it easier to implement drill techniques that maximised firepower. Allied and subsidised units could be expected to fight in an identical fashion with 'national units', and Western imperial powers extended this model to India, training local units to fight as they did. This was a marked contrast to the situation in the Asiatic empires where there were major differences between core and ancillary troops. Nevertheless, Western forces tended to lack a unified political and military command, and effective general staff structures did not develop until the following century.

War of the Bavarian Succession, 1778–9

Prussia's continued high reputation owed something to the brevity of the War of the Bavarian Succession (1778–9), and, more generally, to Frederick the

Great's determination to avoid war and to focus on protecting Prussian interests by diplomacy.[12] The death in 1777 of the childless Elector of Bavaria, Max Joseph III, provided opportunities for Austria to pursue territorial schemes that threatened Frederick's position in Germany. In the resulting war, however, Austria faced Frederick without help from either of its allies, France or Russia; Prussia was also without major allies. In attacking Austria in 1778, Frederick planned the conquest of the key Habsburg dominion of Bohemia by concerted advances from a number of directions. He hoped that diversionary attacks on Moravia and north-east Bohemia would leave the way clear for a march via allied Saxony on Prague by his brother, Prince Henry. However, Frederick's diversionary move into north-east Bohemia was blocked by Austrian fieldworks along the western bank of the upper Elbe. He wanted to breach the Austrian positions near Jaromiersch, but decided that their lines, composed of batteries, palisades and *abates* (ramparts constructed of felled trees) supported by Field Marshal Lacy's 100,000-strong Elbe army, were too strong. Frederick therefore resolved to abandon his original plan and to rely upon Henry's advance to provide a diversion. Henry negotiated a number of supposedly impassable passes, outflanking the Austrians on the middle Elbe, but, instead of pursuing them, he moved to little purpose to the middle Elbe. Affected by supply problems, Henry was also hit by a crisis of confidence. Frederick, meanwhile, quickly crossed the upper Elbe at Hohenelbe, but a rapid Austrian response blocked his advance. Frederick did not attack the Austrian positions, and his campaign ended. Suffering badly from dysentery and desertion, his forces withdrew to Silesia.

In 1779, Frederick proposed an advance on Vienna through Moravia, while the army from Saxony invaded north Bohemia and attacked the line of the upper Elbe from the rear. However, the Austrian army was now far larger than in 1778, while, thanks to the more numerous artillery in both armies, movement was slowed and commanders reluctant to risk attack. Frederick, whose military capacity and energy had clearly diminished since the Seven Years' War, was happy to negotiate peace. By the Treaty of Teschen of 13 May 1779, which was significantly mediated by Russia as well as France, the Austrians failed in their major goal of gaining much territory, but acquired a small and strategic area, the Innviertel.

The war had revealed serious weaknesses in the Prussian army: insufficient supplies, demoralised infantry, undisciplined cavalry, poor medical services and an inadequate artillery train. In some respects, the army Frederick had created was inferior to the one he had inherited from his father and used so successfully in the 1740s. Yet, irrespective of the nature of the Prussian army, Austrian strategy created serious problems for Frederick. The tactical offensive-strategic defensive he had employed so successfully in the Seven Years' War could not be repeated, because the Austrians relied on the tactical defensive-strategic defensive, and did so in an adept fashion, denying Frederick the opportunity to catch advancing forces at a vulnerable moment. Lacy's skill in using massive concentrations of defensive forces in strong positions in the Bohemian hills to thwart Frederick was followed by a post-war fortification policy designed to block possible Prussian invasion routes into Bohemia. This was particularly seen at Theresienstadt, at the confluence of the Elbe and the Eger, and Josephstadt built to complement the pre-war construction of Königgratz on the upper Elbe.

Offering a different focus, in 1778, David Hume, a leading figure in the British Enlightenment, as well as a former diplomat, argued that developments in artillery had led to a lessening of comparative advantage,

> improvements have been continually making on this furious engine, which, though it seemed contrived for the destruction of mankind, and the overthrow of empires, has in the issue rendered battles less bloody, and has given greater stability to civil societies. Nations, by its means, have been brought more to a level: conquests have become less frequent and rapid: success in war has been reduced nearly to be a matter of calculation: And any nation, overmatched by its enemies, either yields to their demands, or secures itself by alliances against their violence and invasion.[13]

The failure to destroy Prussia during the Seven Years' War or to achieve a dramatic, let alone decisive, result in the War of the Bavarian Succession, encouraged such reflections, but they were misleading as a description of warfare on a wider scale in the period. As far as the conflicts in central Europe were concerned, such explanations also suffered from a misleading tendency to search for general rules.

Struggles for control, 1782–92

Although the War of the Bavarian Succession had revealed serious deficiencies in the Prussian military system, there had been no dramatic battle, no equivalent to Rossbach, to register a shift in military prowess. Furthermore, the next combat use of Prussian forces indicated their continued effectiveness, although, in this case, they were greatly helped by the poor quality of the resistance. Frederick the Great's successor, Frederick William II of Prussia (r. 1786–97), intervened on behalf of William V of Orange, who was opposed by the Patriot movement based in the leading cities in Holland. Already, in June, Orangist forces had overcome patriot free corps in Arnhem and Zutphen. On 13 September 1787, 26,000 Prussians, under Charles, Duke of Brunswick, invaded and captured Nijmegen. In the fact of their advance, the principal Patriot force, 7,000 strong under the Rhinegrave of Salm, abandoned Utrecht in panic on the night of 15 September. The Prussians entered Utrecht on 16 September, Gorcum, after a short bombardment, on 17 September, Dordrecht on 18 September and Delft next day. The last Patriot stronghold, Amsterdam, protected by the opening of dykes, held out for longer, but a Prussian force crossed the Haarlem Lake, and the city surrendered on 10 October. In what was an unequal struggle, the absence of an effective Dutch defence had been crucial to rapid Prussian success, while the failure of the French to come to the assistance of their Dutch allies was also important. The French had been greatly affected by domestic political and fiscal problems and were therefore unable to fulfil their threat to intervene, a key instance of the impact of domestic problems on France's ability to sustain a military confrontation.[14]

It was easy to appreciate why it was widely anticipated that Brunswick would have another swift success against revolutionary France in 1792: the Dutch crisis had indicated the effectiveness of Prussian forces and the weakness of a citizens' force. Indeed, Brunswick's failure in 1792 was a product more of command faults and of the extent to which the French were a more numerous and formidable foe than the Dutch forces, than of any particular Prussian deficiency.

Elsewhere in Europe in the 1780s, there was also political strife in Geneva and rebellions against Habsburg rule in the Austrian Netherlands and Hungary. In Geneva, the overthrow of the ruling oligarchy in 1782 was followed by the siege of the city by the forces of France, Berne and Sardinia, and the re-establishment of the old constitution. In Liège, where the bourgeoisie seized power in 1789 during an economic depression, Prussian troops restored the old constitution in 1790. Politics as well as force was crucial, as was seen in Hungary, where the politically astute Leopold II (1790–2) was adept in restoring confidence in Habsburg rule, a marked contrast to the ability of his predecessor, Joseph II, to provoke rebellion.

Eastern Europe, 1787–92

The most active military power in Europe in these years was Russia, and the contrast between Russian operations in 1764–91 and those of Prussia or Austria indicated the extent to which the intensity of struggle was higher in eastern than in central and western Europe. The European crisis touched off by the Turkish declaration of war on Russia in 1787 was to lead also to conflict between Austria and Turkey (1788–90), Sweden and Russia (1788–90) and Sweden and Denmark (1788), as well as to Prussian preparations for war with Austria (1790) and Russia (1791). The crisis also led to the successful Russian invasion of Poland in 1792, which was a prelude to the second partition of that state in 1793, which, in turn, helped lead to a revolt in 1794 that provoked the third, and final, partition of 1795. The warfare of these years was dominated by Russia, whose forces were more successful than those of Joseph II against the Turks, and were able, albeit with some difficulty, to sustain war on two fronts in 1788–90: against the Turks and against Gustavus III of Sweden. In addition, Catherine II was not intimidated when threatened in 1791 by Anglo-Prussian military action in an unsuccessful attempt to force her to accept only modest territorial gains from the Turks. Instead, in the

Ochakov crisis, the British backed down, which led the Prussians to change policy and move towards Russia.

The Turks had begun the war, aiming to seize Kinburn and Kherson, to reconquer the Crimea and to instigate rebellion in the Kuban. There firepower had been crucial in 1783 when the Nogais unsuccessfully resisted being incorporated into Russia. Three thousand of them were killed at the battle of Urai-Ilgasi, in August, by a small, disciplined force under Count Alexander Suvorov; and, on 2 October, in another battle at the confluence of the Kuban and the River Laba, Suvorov again inflicted heavy casualties. The ability to force battles on nomadic and semi-nomadic peoples was crucial to their defeat.

In 1787, a Turkish force landed near Kinburn, but it was defeated by the Russians on 2 October after bitter fighting. Next year, the Russians moved on the attack, focusing on the powerful fortress of Ochakov. Catherine's former lover, Prince Gregory Potemkin, led the besieging army, and bitter naval engagements took place offshore as the Russians struggled to create an effective blockade. After lengthy bombardment, Ochakov was stormed on 17 December. The prevalence of storming in eastern Europe reflected the absence of an equivalent set of rules to those that limited bloodshed in western European sieges. Upstream on the Dniester, Khotin fell to Austro-Russian forces on 19 September after a siege. In 1789, the main Russian army under Potemkin, advanced to the Dniester with naval support, capturing Gadzhibey, Akkerman and Bender. In 1789, Suvorov joined the Austrians in routing the Turks at Fokshani (1 August) and Rymnik (22 September). In 1790, the Turkish navy was defeated in the Black Sea, while the Russians cleared the forts in the Danube delta, capturing Kilia, Tul'cha, Izalchi and Izmail, the last of which was stormed on 22 December. In 1791, the Russians advanced south of the Danube, defeating the Turks at Babadag and Machin. Anara, the Turkish base in the Kuban, was also captured. By the Treaty of Jassy (1792), the Turks recognised the Russian annexation of the Crimea in 1783 and yielded Ochakov and the territory between the rivers Dniester and Bug, a major extension of Russian power on the Black Sea.

The Austrians had less success than their Russian allies. An attempt to surprise Belgrade in early 1788 failed, the Austrians losing their way in the fog. Deploying 140,000 troops, Joseph II hoped to conquer Serbia, Moldavia, Wallachia and most of Bosnia, but the Turks concentrated their efforts against the Austrians and not against the Russians. Joseph proved an indecisive commander-in-chief, disease debilitated the Austrians, and the Turks were able to put them on to the defensive. Invading the Banat of Temesvár in August 1788, the Turks, under Grand Vizier Yussif Pasha, defeated the Austrians in an attack on their camp at Slatina, from which Joseph only narrowly escaped. The Turks were finally halted by their own supply problems, rather than by Austrian resistance. In 1789, the Austrians were more successful. In place of the linear tactics they had adopted against the Turks in 1737–9, they now used flexible infantry squares arranged to offer mutual support. This brought victory at the Battle of Mehadia (23 August),[15] which was followed by the crossing of the River Sava and the siege of Belgrade. The Turkish defeat at Rymnik lessened the chance of relief, and, after a massive bombardment, Belgrade surrendered in October. Bucharest fell soon afterwards and the Austrians occupied Serbia, although, faced by serious domestic problems, particularly in Hungary, and by Prussian pressure, Leopold returned these gains when the Treaty of Sistova was concluded in 1791.

Fighting in northern Europe was on a much smaller scale. As a result of Anglo-Prussian pressure, Denmark swiftly ended its attack on Sweden in 1788 (which was designed to benefit from the latter's conflict with Denmark's ally Russia). Furthermore, the 10,000 strong Norwegian force that marched on Gothenburg suffered losses of 3,500 men that winter due to disease. The war between Russia and Sweden was longer, and, in a series of hard-fought struggles Gustavus III of Sweden was thwarted. He invaded Russian Finland in 1788, bombarded Fredrikshamn and unsuccessfully besieged Nyshott, before retreating in the face of political hostility on the part of his own officers. At sea, thanks to new types of vessel developed in the 1760s and 1770s, especially versatile oared gunboats, as well as ships of the line built in the 1780s, Sweden was a formidable challenge. However, a drawn battle off Hogland in the Gulf of Finland on 17 July 1788, in which the Swedes were hindered by ammunition shortages, denied Gustavus the crucial control of the Gulf which he needed both for his military operations in Finland and if he was to carry out an amphibious attack on the Russian capital, St Petersburg. His attempts to seize the city in 1790 was similarly unsuccessful, attacks on Russian squadrons failing in part because the Swedes lacked tactical training in offensive warfare.[16] Eventually, the Russians were defeated in a major engagement between the two archipelago (oared) fleets, but the cumulative strain of recent battles on the Swedish fleet was considerable, both sides had now exhausted their abilities for further offensive warfare during 1790 and, crucially, Britain and Prussia had not attacked Russia as Gustavus had hoped. He felt obliged to negotiate the Peace of Verela (1790) without making any territorial gains. This conflict was far distant from that which the British waged in southern India in 1790–2, the Third Mysore War, but they indicated the range and diversity of the military system about to be assaulted by Revolutionary France.

9

WARFARE IN THE FRENCH REVOLUTIONARY AND NAPOLEONIC ERA, 1792–1815

> I will not remind those gentlemen of their declaration, so often made, that the French must fly before troops well disciplined and regularly paid. We have fatal experience of the folly of those declarations; we have seen soldiers frequently without pay, and without sufficient provisions, put to rout the best-paid armies in Europe.
>
> British opposition spokesman (and playwright), Richard Brinsley Sheridan, House of Commons, 4 April 1797.[1]

Sheridan was correct, although as we shall see, there are important qualifications that must be made. Under her Revolutionary governments, French forces had more success in Europe than their predecessors since the 1690s, although this success was conspicuously not matched at sea. In part, this was a reflection of British strength in this sphere and of British concentration on war with France, whereas, on land, Austria, and, even more, Prussia and Russia, also had other interests to pursue (particularly in Poland), a reminder of the primacy of political considerations. Other factors, however, were also pertinent, although it is difficult to gauge their respective impact. At sea, the French could not match the organisational and tactical advances they showed on land. In part, this was because naval capability and warfare were more reliant on institutional continuity. In contrast, the adversarial nature of the French Revolution and its commitment to struggle helped lead to a major increase in the French resources available for war on land, although campaigning swiftly became dependent on the looting of conquered areas.

The French Revolutionary War

The French Revolutionary War began in 1792 with conflict between France and an alliance of Austria and Prussia. A mutual lack of sympathy and understanding between Austria and France, and a shared conviction that the other power was weak and would yield to intimidation helped ensure that war with Austria broke out in April 1792. France declared war on Austria in the hope that a revolutionary crusade would defeat the enemies of the Revolution, consolidating it in France and extending the revolutionary example elsewhere in Europe. A variety of proposals for the reorganisation of Europe was advanced, including the extension of France to her 'natural limits' and the division of much of Europe into small republics. Refugees from failed revolutions elsewhere in Europe, particularly from the Austrian Netherlands and the prince-bishopric of Liège, as well as Dutch 'Patriots' and radicals from other countries, such as Britain, encouraged deceptive opinions concerning the willingness of their compatriots to rebel and to help France and pressed strongly for French assistance in remodelling the Low Countries. In contrast, Frederick William II of Prussia declared war on France in May 1792. The war further broadened out the following year, when Britain, Spain and the Dutch joined the conflict, so that it involved more of western Europe in war than at any time since the War of the Austrian Succession in the 1740s.

This was an unprecedented challenge to France, which had, during the 1740s, been allied with Spain. However, the *levée en masse*, a general conscription for all single men between the ages of eighteen and twenty-five, ordered in August 1793, raised large forces, such that the French were able to operate effectively on several fronts at once, to sustain heavy casualties and to match the opposing forces of much of Europe. The distinction between the volunteer professional army and the conscript militia had been ended. Such powers of conscription were not new in Europe and, particularly as a result of draft avoidance and desertion, it anyway proved difficult in France and areas annexed by France to raise the numbers that had been anticipated. The French war effort, nevertheless, owed much to the combination of a large population and a mobilisation of resources made possible by the extension of government power. In 1793–4 alone, nearly 7,000 new cannon and howitzers were cast by the French. Although there were major difficulties, especially in logistics, the French army was successfully moulded and sustained as a war-winning force. Indeed, the *levée en masse*, the boldness of the patriotic vision it represented and the numbers it raised came to play a key role in the mythology of warfare, appealing alike to imaginative nationalists and determined warleaders,[2] although, as more generally with revolutionary armies, aspects of the myth, particularly the claims to novelty and potency, were misleading.[3]

The French had mass and system. The new logistics brought about by the partial abandonment of the magazine system helped the aggressive style of war – both in strategy and in tactics – of revolutionary armies, able to rely on numbers and enthusiasm. The political context provided a frenetic energy to the conduct of war by France. It has been argued that the French soldiers were better motivated, and hence more successful and better able to use the new methods. Revolutionary zeal had greatly declined by 1797, but, initially at least, revolutionary enthusiasm does seem to have been an important element in French capability.[4] It was probably necessary to counter the effects of the limited training of many of the revolutionary soldiers and for the greater morale needed for effective shock action, particularly for crossing the killing ground produced by opposing fire. The ready recourse to violence in war was, in part, a product of the role of civil violence, anxiety and a politics of paranoia in the early years of the Revolution,[5] and this had artistic echoes. In his painting *Marius Returning to Rome* (1789), Baron François-Pascal-Simon Gérard had prefigured the iron determination of revolutionary violence when he showed a demonic Marius leading a column of troops of republican Rome with the heads of their victims on spears. Civilians are being slaughtered and there is terror on the face of the people.

Waging war by the brutalisation of subjects and the despoliation of foreigners produced results. The exploitative nature of French rule abroad led to a crucial increase in resources which complemented France's domestic revolution, although the exploitation helped to limit the popularity of the revolution outside France, a process accentuated by the length of the conflict and the size of the French forces. This encouraged rebellion, and at a large-scale level, and also created within France an atmosphere of expediency in which it became easier for a general to reach for power. Indeed, helped by local hostility to the radicalism and expropriations of the French conquerors, the Austrians regained the Austrian Netherlands after their victory at Neerwinden (18 March 1793); but, by the end of 1794, having defeated the Austrians at Fleurus (26 June 1794), the French had again conquered it, before going on to overrun the United Provinces (Netherlands) as Austrian forces fell back to the Rhine. British and Dutch troops were unable to stem the French.

Amsterdam fell on 20 January 1795, a target that had eluded Louis XIV in 1672. Convinced that France could not be defeated, Prussia and Spain made peace that year. Prussia and, even more, Russia were more concerned about developments in Poland, which was partitioned out of existence in 1795.[6] This concern with Poland helped ensure a lessening of the focus on France throughout the early stages of the French Revolutionary war, and Russian forces were not effectively deployed against France until 1799.

The superiority in French numbers was important both in individual battles, such as Valmy (20 September 1792), Jemappes (6 November 1792) and Wattignies (15–16 October 1793), and in offensives, such as that in which the Spaniards were driven out of Roussillon in 1794. At Valmy, the larger French force blocked the Prussian advance on Paris, ensuring that there would be no repetition of the Duke of Brunswick's successful invasion of the United Provinces in 1787. The French could afford their greater losses in battles such as Jemappes and Fleurus.

Tactics were also important. The characteristic battlefield manoeuvre of French Revolutionary forces, and the most effective way to use the mass of new and inexperienced soldiers, most of whom went into the infantry, was the advance in independent attack columns. This was best for an army that put an emphasis on the attack. It was easier to train hastily raised troops to fight in columns. Little time was required, and the formation offered a psychological benefit of packing the recruit in with others, establishing a sense of security in the midst of battle. Column attacks were particularly successful against the Austrian linear defensive formation.

It is necessary, however, to distinguish between shallow columns and deep columns, and between columns of manoeuvre and columns of attack. The French Revolutionary armies did not apply these principles in practice zealously, for levels of training suffered in the early 1790s, especially with the new armies, and everything tended to break down when combat began; but there is a need for a distinction between the image of some enormously broad and deep formation, with men packed tightly together, and what was probably more common, a less coherent and dense formation. There was a difference between battalion columns and brigade/division columns: the former was a single block of men, say fifty wide by twelve deep, whereas the latter could be made up in many different ways, with variations on possible formations achieved by varying the spacing of the individual battalions and lines of battalions. Nevertheless, in general, column attacks reflected and required substantial numbers of troops. Some French columns were 5,000 men strong.

The French used concentration at two levels: tactical and operational. In the first, their opponents were vulnerable because of an emphasis on traditional close-order linear formations, while the French system allowed for more men to be concentrated in a smaller area than before. At operational level, the French struck at the Austrian 'cordon system' which distributed units to hold territory and so reduced the numbers that could be concentrated at any one point. The large number of separate armies that the French deployed needed substantial numbers of troops, as did the various divisions of individual armies. The former were required so that France could operate on the series of fronts that the number of her opponents required, while the latter were necessary for the strategy of envelopment, encirclement and convergence that was so important for Revolutionary and Napoleonic warfare. The greater number and dispersal of units ensured that command and coordination skills became more important, although the introduction of the corps-division-brigade structure countered the increase in the number of units and their dispersal.

Column advances were more flexible than traditional linear formations and rigid drill. Furthermore, as their units became more experienced, the supply of equipment improved and vigorous generals rose to the fore, the French forces were steadily more effective. In tactical terms, they were helped by a turn towards light infantry, so that a substantial force of skirmishers, using individually aimed fire, weakened the opposing lines before the French attack. An ability to move quickly between line and column formations also proved important.

Deployed in open order, the skirmishers were not vulnerable to volley fire or cannon, but were able to use individual fire in order to inflict casualties on the close-packed lines of their opponents. This hit enemy morale and could also disrupt their formations. Other armies had previously used light infantry in a largely separate capacity on the battlefield, but the French integrated close-order and open-order infantry, so that every infantry battalion had the capacity to deploy its own skirmishers. However, as with the use of rifles by the American Revolutionaries, it is important not to exaggerate the impact of a particular weapon, tactic or deployment. None was unique to the French (the Russians being expert in the use of columns), and each faced particular problems. If the advance of the head of a column was stopped by defensive fire, this could have a disastrous consequence. Furthermore, skirmishers were vulnerable to cavalry, although far less so in broken terrain, and they were also used by the Allies, although not in the way used by the French.

The impressive French battlefield artillery was also important. It was an inheritance from the pre-Revolutionary army but was taken forward with the development of horse artillery from 1791. This provided close support for the infantry. The French combination of mobile artillery, skirmishers and assault columns was potent, an original and disconcerting ad-hoc combination of tactical elements matched to the technology of the time and the character of the new republican soldier. Pre-Revolutionary military thought (see pp. 110–11) had an impact.

At the same time, the course of individual battles indicated that notions of French superiority need to be qualified. At Jemappes (6 November 1792), the battle that led to the initial Austrian loss of the Austrian Netherlands (Belgium), the French fielded 38,000 troops and 100 cannon, and the Austrians only 13,200 troops and fifty-four cannon. Revolutionary enthusiasm may well have played a role in encouraging the repeated assaults by the thirty-eight battalions of French volunteers, but there were also thirty-two battalions of regular infantry, still wearing their white Bourbon uniforms. Both groups had benefited from the intensive training to which Dumouriez had subjected his forces in the summer of 1792. Far from enjoying clear battlefield superiority, however, the initial French attacks were beaten back by defensive fire. Eventually, the pressure of greater French numbers, and the threat of being outflanked, led the Austrians to retreat, but they had held out for four hours, escaped successfully, and had 1,241 casualties, compared to 4,000 from the French. Nevertheless, thanks to Jemappes (although also to the demolition of fortifications in the 1780s as Joseph II got rid of the Dutch-garrisoned Barrier forts), Belgium fell far more rapidly than when attacked by the French last, in 1744–7. The Austrians retreated towards the Rhineland. Brussels fell on 13 November and Ostend three days later. Only isolated posts, principally the citadels in Antwerp and Namur, were left. They surrendered on 29 November and 2 December respectively.

The campaign of 1792 demonstrated a number of features of European warfare that were to be characteristic of the period up to 1815. Battles were generally more important than sieges. For example, after their advance on Paris was checked at Valmy, the Prussians under the Duke of Brunswick retreated, ending the siege of Thionville and abandoning the fortified positions of

Longwy and Verdun. Nevertheless, the career of Napoleon showed, most clearly at Toulon (1793), Mantua (1796–7), and Acre (1799), that sieges could still be as important as battles. Success at Toulon made his name, Mantua showed that sieges could lead to relief attempts that caused important battles, and the inability to capture Acre led to the failure of his Palestine campaign. Clearly, much depended upon the particular theatre of operations. Fortified positions and sieges still remained important in the Peninsular War, and indeed Wellington's creation of the Lines of Torres Vedras to protect Lisbon and his army led to the establishment of a large-scale entrenched position that thwarted the French advance in 1810–11. However, sieges played a smaller role in conflict in central and eastern Europe, although even there they could be important, as with the successful French siege of Prussian-held Danzig in 1807. In 1813–14, fortresses played a major role in operations and strategy. Napoleon used Dresden as a key *point d'appui* in 1813, while he insisted on trying to hold onto fortresses in East Prussia, Silesia, Poland and Catalonia-Valencia.

Speed was another feature of campaigning, especially of French advances from 1792, for example the successful invasion of Savoy that year, and Napoleon was a beneficiary, not an initiator, of this change. The battlefield use of artillery, by both attacking and defending armies, was of great importance. There was an increased use and massing of artillery on battlefields, and artillery was beginning to dominate battlefields in a way it had not done before. Ruthless cavalry follow-ups after victory were also important in consolidating success, as after Napoleon's victory over the Prussians at Jena (1806), while, conversely, his lack of cavalry after the unsuccessful invasion of Russia in 1812, was partly responsible for the failure to exploit his victories in 1813, for example Lützen.

The politics of revolution ensured that systems of command in France in the 1790s differed from those of the *ancien régime*. There was a more 'democratic' command structure, at least at battalion level: the social gap between non-commissioned officers and their superiors was less than hitherto. The French also benefited from young, energetic and determined commanders. The Revolution created an officer class dominated by talent and connections, as opposed to birth and connections, although the election of officers was rapidly abandoned. The emphasis on republican virtue, seen in 1792–4, was replaced by an emphasis on professionalism, merit no longer being regarded, as it had been in 1792–4, as a dangerous sign of individualism.[7]

Despite these shifts, and the major restrictions created by political associations, careers were open to talent, commanders including Jean-Baptiste Jourdan, a former private; Louis Lazare Hoche, a former corporal; and Pichegru, a sergeant major in the artillery from a peasant background. Napoleon Bonaparte, initially a junior artillery officer from the recent acquisition of Corsica, was an instance of the petty nobility that had been the backbone of the French army throughout the eighteenth century. Commanders who failed, or were suspected of treachery, were executed, and indeed there were some traitors. In March 1793, Charles Dumouriez, commander of the Army of the North, concluded agreements with Austrian commanders which were intended to allow him to suppress the Revolution in Paris, but he was thwarted. The use of revolutionary violence, nevertheless, lacked discrimination. Eight days after the outbreak of war, one French army murdered its unsuccessful commander, Théobald Dillon. Other generals followed, including seventeen killed in 1793 and sixty-seven in 1794. Houchard was even executed for achieving only moderate success in 1793. Punishment and politicisation helped ensure that in battle and on campaign the generals were willing to accept heavy losses among their troops.

Initial confusion in military administration was followed by a measure of organisation, as the government struggled to equip, train, feed and control the new armies. This owed much to Lazare Carnot, head of the military section of the Committee of Public Safety, and, like Napoleon, a product of the reformed royal army of the years prior to the Revolution. He improved the equipment, artillery and staff organisation of the armies, ended the election of officers and melded together veteran and new units, not least in the adoption of the three-battalion regimental system in early 1794. War led to organisational development,[8] including that of the all-arms division and, in contrast, that of the consolidation of army artillery forces, but such development was affected by differences of opinion. There was also administrative rationalisation, although the process was heavily politicised, not a matter of abstract bureaucratic reform. Nevertheless, the heady intoxication of the new, and of its apparent benefits,[9] prefigured that of the impact of flight in the early twentieth century. Indeed, the French army formed the world's first air force, even if it was only equipped with hot air balloons.

The process of forming the new armies and using them with success was instrumental in the transition from a royal army to the nation in arms. The way was open for the ruthless boldness that Napoleon was to show in Italy in 1796–7. Alongside the emphasis on change and development, however, it has been argued that French generals did not practice 'revolutionary' conflict, but, rather, were effective exponents of *ancien régime* warfare, especially of 'all the many short-cuts ... that the Old Regime had itself already authorised.'[10] The degree to which French war-making built on pre-Revolutionary developments, alongside the resilience of Austrian forces in the 1790s, indeed indicates that the French Revolution did not make redundant what had come earlier, a situation even more clearly exemplified by continued British success at sea. The Prussians also fought well in the early 1790s, proving adept both as light infantry and in line formations.[11] On land, the successful British army of the Peninsular War (1808–14) and Waterloo (1815), was very much an eighteenth-century-type army, albeit one improved in a deliberate process in the 1790s and 1800s.[12] The Austrians, however, showed that there was a limit to what they could do to emulate Revolutionary and Napoleonic France.[13] Similarly, in Spain, there was an attempt to reform the army, although great obstacles were faced.

Force was also employed in France in dealing with internal opponents of the revolution, the army overlapping in this sphere with other institutions of revolutionary violence. In 1791, it had been hoped that the army could be used to stabilise the revolution, but disorder within it and uncertain direction had helped thwart this policy,[14] rather as they had prevented its use for counter-revolutionary purposes in 1789. Direction proved more clear-cut in 1793–4. Royalists were a target for action, especially in western France, where a rising in the Vendée in 1793 was triggered by attempts to enforce the *levée en masse*. Initial royalist success, which benefited from the advantages of fighting in wooded terrain, led to brutal repression, including widespread atrocities against non-combatants, as well as the destruction of crops seen in some transoceanic Western conflict with natives. More generally, the army served as a coercive agency of de-Christianism, enforcing the ban on Christian practice. In contrast, strong piety was linked to more explicitly anti-revolutionary violence and to support for royalism. The rebels in the Vendée called themselves the Royal and Catholic Army, and the atrocities of the republican forces spurred them to fresh activity. The Battle of Cholet was one of the largest in the French revolutionary wars, but after December 1793 the rebels concentrated on partisan action. However, the switch by the government to a more conciliatory stance led some of the rebels to agree terms in January 1794. The need for a political dimension as well as a military solution was captured in Hoche's instructions to the troops fighting the Chouans, written in 1794–5,

> Whilst swearing to wage a war to the death against those miscreants who have refused to profit from the National Convention's amnesty, the Republican troops will, however, respect the peaceful inhabitants of the region. This will enable them to distinguish between Republicans doing their duty and those detestable individuals who have chosen to follow the despicable career of robbers and murderers.[15]

This distinction proved difficult to make, not least because of the demonisation of opponents indicated by the last comment.

Provincial opposition was not only mounted by royalists. The division among the revolutionaries that resulted in the purge of the Girondins by the Jacobins, led, in 1793, to a series of revolts, especially in southern France, termed 'federalist' by the revolutionaries. Those in Bordeaux and Marseille were swiftly repressed, but the opposition in Lyon and Toulon was fierce. These revolts overlapped with counter-revolutionary activity, and much of western and southern France was in a state at least close to insurrection from 1792. Particular disturbances arose from a background of widespread instability.

Radical governmental policies encouraged popular opposition, but, in 1794, there was a reaction at the centre against radicalism. Following the coup of 9 Thermidor (named from the month in the revolutionary calendar), the government, instead, became the force of propertied order. As such, it was willing and able to defeat radical and royalist popular uprisings in 1795. Continued foreign war, however, prevented any real stability in France. The new Directory government believed foreign war necessary in order to support the army, to please its generals and, for these and other reasons, to control discontent in France, not least by providing occupation for the volatile generals, whom it found it difficult to manage. The views and ambitions of many of the generals were not limited to the conduct of war, and many were contemptuous of civilian authority and involved in politics: Pichegru, Hoche and Napoleon played roles in the 1797 Fructidor coup. This role was an aspect of a more widespread collapse of discipline also seen in large-scale pillaging by troops.

Furthermore, victories abroad led to pressure for further conquest, in order to satisfy political and military ambitions and exigencies. The French, however, did not push all before them. The Austrians proved tough opponents, especially in the Austrian Netherlands in 1793 and, under Archduke Charles, in Germany in 1796. Disciplined Austrian fire initially checked the French at Jemappes and Hondschoote (8 September 1793), and helped to defeat them at Neerwinden and Weissenburg (13 October 1793). In 1795, the French in Germany were pushed back across the Rhine: Würmser and Clerfayt successively defeated Pichegru, and Würmser recaptured Mannheim. In 1796, Archduke Charles beat the French at Wetzlar (15 June), Amberg (24 August) and Würzburg (3 September), Amberg stopping a French advance on Austria.[16] The Prussians, under the Duke of Brunswick, had won victories at Pirasens (14 September 1793) and Kaiserlauten (28–30 November 1793). Before leaving the First Coalition on its own terms because of growing concern over Russian interests in Poland, the Prussians had fought the French to a standstill on the central and lower Rhine. The Prussians perceived their performance during the Revolutionary War as quite successful and continued to see it as such until 1806. The Russians were to show impressive staying power and fighting quality, both in 1799 and subsequently.

However, Napoleon, the new commander of the French army in Italy, developed in 1796 the characteristics of his generalship which was to prove so important in establishing France's military reputation: self-confidence, swift decision-making, rapid mobility, the concentration of strength and, where possible, the exploitation of interior lines, achieving, as a result, the outcome that French forces on this front had failed to secure since the outset of the war in 1792. Victory at Mondovi (27 April), by French columns over outnumbered defenders, knocked Sardinia out of the war (it was then forced to support France militarily in 1797), while at Lodi (10 May), Bassano (8 September), Arcola (15–17 November 1796) and Rivoli (14 January 1797), Napoleon's ruthless boldness and ability to manoeuvre on the battlefield brought victory over the Austrians and associated Napoleon with military success; meanwhile, in 1796, the destruction by his forces of a popular revolt at Pavia served both as an assurance of his revolutionary vigour and a reminder that this was a conflict of rival cultures as well as rival politics. The destruction of the Pavian Revolt, which entailed summary executions and the burning of villages, suggests that popular resistance could do little to hinder the French, although popular opposition made progress difficult for the French in central Italy. Austria left the war by the Treaty of Campo Formio (1797), under which it ceded Belgium and Lombardy.

The Directory, however, could not free France from the burdens of war, while the political-military structures it had created in occupied territories were unstable and faced by resistance, including repeated rebellion, as in the Roman Republic in 1798. 'It is absolutely the Vendée over again', declared one French general.[17] The Second Coalition (of powers opposed to France), formed in 1798–9, put major strain on France. In November 1798, the Neapolitan army invaded the Roman Republic. The Austrians under Archduke Charles defeated a smaller French army at Stockach (25 March 1799), while a Russian army under Alexander Suvorov advanced into northern Italy, the first time that the Russians had operated there. Suvorov's victories, especially Trebbia (17–19 June) and Novi (15 August 1799), were brutal battles in which repeated attacks finally found weaknesses in the French position. Like Napoleon, Suvorov was a believer in the operational and tactical offensive and had little time for sieges. He was also willing to accept a high rate of casualties and to mount costly frontal attacks on fortified positions, such as the castle of Novi or the northern approaches to the St Gotthard Pass. Suvorov relied on bayonet attacks rather than on the use of defensive firepower. Suvorov's generalship contrasted with that of his more cautious Austrian allies, who were more concerned with regaining northern Italian positions by sieges. They complained to Suvorov and to Tsar Paul that the marshal's tactics were too costly, that he favoured assault to firepower, and that Austrian regiments were earmarked by him for these assaults, while Russian regiments were not always thrown into the fire. Suvorov's generalship, both tactical and operational, demonstrated clearly that a determination to pursue military goals aggressively, and at the risk of high casualties, was not restricted to the Revolutionary French.[18]

The wars of Napoleon

The French home base was not overthrown in the War of the Second Coalition, but the discredited government was replaced in November 1799 by a coup in which Napoleon was the key figure. His rise to power, which, in 1804, culminated when he crowned himself Emperor, marked the end of domestic radicalism. Abroad, Napoleon showed a singular unwillingness to accept compromise, a desire, at once opportunistic, brutal and modernising, to remould Europe, a cynical exploitation of allies, and a ruthless reliance on the politics of expropriation, which prevented international stability. Thanks largely to Napoleon, the nineteenth century began under the shadow of war.

Napoleon was in a position not only to act as an innovative general but also to direct the French war effort. Resources were devoted to the military with a consistency that the Revolutionary governments had lacked, although his determination to retain control over military administration created problems for the latter not least when he was on campaign in 1800. As a result, Napoleon pushed through an administrative reordering in 1801–2 that produced a more effective system that was also acceptable to him as it did not pose a political challenge.[19] The conscription system, which had become less effective in the late 1790s, was strengthened, while Napoleon also benefited from the accumulated experience of an army many of whom were professionals for whom campaigning was not only their life, but also one whose problems they had learned to master.[20]

In a key organisational move, that gave the French an important comparative advantage,[21] Napoleon also developed the corps in 1800–1, as a level above that of the division which could include all the arms and also be large enough to operate effectively: both corps and divisions were given effective staff structures. Thus the corps added, to the flexibility of the earlier divisional system, the strength necessary both for the grinding, if not attritional, battles of the period, where opposing forces would not collapse rapidly as a result of well-planned battlefield moves, and for Napoleon's campaigns of strategically applied force. Corps operated effectively both as individual units, as at the Battle of Auerstädt (1806), and in concert, and they helped make what is now referred to as operational warfare more feasible. It is important not to idealise organisational units and developments, as there was considerable variety in practice, but the corps system nevertheless enhanced military strength, not least by enabling the different arms to be used more effectively. In the 1790s, the French division had been a force of all arms, for example two brigades of infantry, one of cavalry and a battery of guns. This, however, meant that the artillery and cavalry were split up into 'penny packets' and deprived of much of their striking power. Under the corps system, by contrast, both cavalry and infantry tended to be controlled at a higher level and even grouped into separate formations altogether, for example artillery reserve and cavalry corps, providing an important force multiplier. This, however, risked creating fresh problems, as in the Franco-Prussian War of 1870–1 when the French artillery reserves were not in the right place at the right time.

Skilled staff work was important to Napoleon as it resulted in the movement by different routes of several individual corps which were still able to support each other. His chief of staff, Louis-Alexandre Berthier, was a crucial figure. Napoleon thus had a better command structure than his opponents, a contrast that helped him to victory over the Austrians in 1805, although his centralised direction of campaigns became a problem when he limited the autonomy of his commanders in the distant Peninsular War (1808–14). Moreover, it proved difficult to control the large armies of his later years which were under his direct command.

The French organisational and command structures were vital to Napoleon's characteristic rapidity of operational and tactical movement, while his troops also travelled more lightly than those of Frederick the Great. Napoleon exploited this mobility to operational and strategic effect. The manoeuvre of the central position was employed to divide more numerous opposing forces and then to defeat them separately; while envelopment was used against weaker forces: they were pinned down by an attack mounted by a section of the French army, while most of the army enveloped them by cleaving past them and then cutting their lines of supply. Napoleon's determination to destroy the opposing army has been contrasted with a more cautious and conservative command attitude and style on the part of Archduke Charles.

Napoleon sought battle. On the battlefield, he was a firm believer in the efficacy of artillery, organised into powerful batteries, especially of 12-pounders, the production of which was pressed forward in 1805 in pursuit of the advice of a committee under General Marmont which had met in 1802–3 and which recommended, as part of the system of Year IX, the replacement of 8-pounders by the heavier 12-pounders. Napoleon's experience of the devastating massed fire of the more heavily gunned Russians at Eylau (1807) led him to press forward his own massing of artillery, greatly increasing the number of cannon in his massed batteries.[22] At Wagram (1809), he covered the reorganisation of his attack with a battery of 112 guns, at Borodino (1812) about 200 cannon were massed, while at Lützen (1813) the artillery played a major role in weakening the Prussian centre. Napoleon also increased the amount of shot available per cannon in order to make continuous fire possible. Conversely, at Aspern-Essling (1809), an Austrian battery of 200 guns had inflicted serious damage on the French, this a reflection of Austrian improvements from 1805. Napoleonic warfare thus looked towards the heavy use of artillery in subsequent Western warfare.

The French heavy cavalry was similarly massed for use at the vital moment, as with Murat's charge through the Russian centre at Eylau (1807). For the infantry, Napoleon used *l'ordre mixte:* a mixture of lines and columns with many sharpshooters to precede the attack. This was a potent organisation, but French tactics degenerated later on during Napoleon's reign, and he devoted insufficient attention to tactical details.

As a commander, Napoleon was similar to the British admiral, Horatio Nelson. Both were dominated by the desire to engage and win, although they could make operational blunders, as Nelson did when he missed Napoleon en route to Egypt in 1798 and when he fruitlessly chased the French admiral Villeneuve to the West Indies in 1805. By forcing a battle whose shape was unclear, both Napoleon and Nelson placed great reliance on the subsequent mêlée, which rewarded the fighting qualities of individual units, the initiative and skill of subordinates, and, in Napoleon's case, the ability to retain reserves until the crucial moment.[23] Both deserve high praise for having prepared the effective military machines that their victories revealed.

Napoleon's opening campaign as First Consul was an invasion of northern Italy boldly begun with a crossing of the Great St Bernard Pass so that he arrived in the Austrian rear. At Marengo (14 June 1800), however, he found the Austrians to be a formidable rival, and Napoleon's enforced retreat for much of the battle was only reversed because of a successful counter-attack mounted by French reinforcements. Nevertheless, victory it was and, thanks in part to his favourable spin on his generalship, Napoleon's grasp of power was cemented. At Hohenlinden (3 December 1800), east of Munich, flexibility in defence helped Jean Moreau defeat the Austrians, and peace was concluded the following February at Lunéville.

Conflict with Austria resumed in 1805, as the War of the Third Coalition began. Napoleon moved the Grande Armée (originally deployed at Boulogne in preparation for an invasion of England) into Germany by rapid marches with supplies crucially provided by German rulers who Napoleon both wooed and intimidated.[24] The Bavarians also provided Napoleon with troops. The advance of separate corps across a broad front lessened the logistical strain and increased the speed of advance. Napoleon outmanoeuvred the Austrian army under Karl von Mack, based at Ulm and preparing for an attack from the west through the Black Forest and not for the advance from the middle Rhine to the Danube in the Austrian rear which the French achieved. The surrender of the Austrians at Ulm on 20 October 1805 was followed by the rapid overrunning of southern Germany and Austria, Vienna being occupied as an open city on 12 November. This undercut Austrian operations in Italy, where their largest army was deployed under Archduke Charles to little effect.[25]

The Russian forces sent to help Austria concentrated in Moravia, and Tsar Alexander I, overconfident about his strength, decided to attack Napoleon at Austerlitz on 2 December 1805. A strong Russian attack on Napoleon's right was held in marshy terrain by French infantry, and the French then turned the weak flank of this attacking force to crush the Russians. The French had 8,000 men killed or wounded, the Austrians and Russians 16,000, and about 11,000 were taken prisoner. Defeat led Austria to accept a harsh peace. The following September, Frederick William III of Prussia joined Russia, but Napoleon attacked him before Russian reinforcements could arrive. At the battles of Jena and Auerstädt (14 October 1806), the poorly commanded Prussian forces were defeated. Napoleon crushed what he thought was the main Prussian army at Jena, massed artillery and substantial numbers of skirmishers inflicting heavy losses on the Prussian lines; while, at Auerstädt, Davout, with the 27,000 men of III Corps, held off and finally defeated the main Prussian army of at least 50,000 men. These victories encouraged Selim III of Turkey to respond to the Russian occupation of Moldavia in November by declaring war, as Napoleon wished, rather than yielding to Russian pressure.

Napoleon's successes in 1805–6 were very different to those of the French Revolutionary forces. There was no equivalent in the 1790s to Napoleon's successful advance into central Europe, nor to the defeat of the main Prussian and Russian armies. However, to judge the Revolutionaries by pointing out that Napoleon achieved more far-flung and rapid triumphs is, in part, harsh.

The situation facing the French was more difficult in the 1790s than it was in the 1800s, and Napoleon's military position was a creation of the 1790s, indeed, a solution to the problems it posed.

In 1807, the Russians were engaged at Eylau in East Prussia (8 February). Russian attacks pressed the French hard; and repeated French attacks failed to break the Russians, who withdrew during the night. French casualties were heavy (more than Austerlitz, Jena and Auerstädt combined) and, although Napoleon had gained possession of the battlefield, and Russian losses were heavier, he had won neither tactically nor operationally. At nearby Friedland, however, on 14 June, the Russians attacked with an inferior force and with their back to a river, losing heavily. On 7 July, the Treaty of Tilsit ended hostilities between France and Russia on French terms, although they were very generous. Friedland had left the Russians so battered that they needed time to recoup losses and to rebuild the army. Frederick William III was obliged to accept another Treaty of Tilsit with major territorial losses, which was followed by the Treaty of Paris of September 1808, which restricted the size of his army to 42,000 men and specified a heavy indemnity. The prohibition of conscription, as well as of a militia and a national guard, was designed to keep Prussia weak. Prussian territorial losses, like the occupation of Hanover, provided Napoleon with the opportunity to remodel Germany to the benefit of allies, such as Bavaria, and members of his own family, who were given territories.

In November 1807, the French invaded Portugal successfully, but Napoleon's attempt to seize Spain led to a popular uprising in 1808. A surrounded French corps of 20,000 men surrendered at Bailen on 21 July, and the new king of Spain, Napoleon's brother Joseph, retreated from Madrid. This was the most dramatic insurrection in the period, but it was by no means the only one. Revolutionary French forces and their allies, especially if, as in Italy, they were anti-clerical, had met vigorous popular resistance, particularly the Sanfedesti Crusade under Cardinal Fabrizio Ruffo in the kingdom of Naples in 1799, which led to the creation of a peasant army of possibly 100,000 men. French rule also encountered widespread resistance which was less conspicuous. For example, in the summer of 1793, an insurrection in the mountains near Delémont in the prince-bishopric of Basle, which had recently been annexed to France, received no outside support and was defeated by overwhelming French forces. In 1798, a rising against French rule in Cairo was quelled by French shelling and, in 1800, a more sustained rising was finally coerced into surrender. In contrast, the Calabrian rising of 1806 was initially helped by the British, as well as by the distance from centres of French power, and the logistical problems facing the French which led to them seeking food in a region where it was sparse. It took a considerable period and much effort, including the use of flying columns, to suppress it finally, although a lack of effective rebel leadership, combined with the French ability to exploit local vendettas, both helped in the eventual outcome.[26]

Some of the popular forces that emerged had a more tangential bearing on the struggle between France and her opponents. In 1798, an Irish nationalist rising against British rule was quickly suppressed, French intervention on behalf of the rebels proving unsuccessful; and in 1808 a *levée en masse* was decreed for the defence of Sweden against Russia, France and Denmark. The Serbian uprising of 1804–13 began as a reaction to harsh control by semi-independent Turkish troops and became a fight for independence. As in other popular risings of the period, there were social and ethnic dimensions to the conflict. The Serbs plundered and destroyed Muslim settlements and drove the Muslims out, but they were defeated by Turkish forces who, in turn, burned down Serbian villages.[27] The Serbs were unfortunate in the international context, with Russia ending its war with Turkey in 1812 and earlier focusing on conflict in the lower Danube, while Austria was absorbed by the politics of responding to Napoleon. In 1814, there was opposition in Norway to its transfer from Danish and Swedish rule. Invading Swedish troops were defeated at Lier and Skotterud, skirmishes which were Norwegian envelopment victories, but a Norwegian realisation of the strength of the experienced Swedish military and the willingness of Karl Johan of Sweden in the negotiations which continued during the fighting – two aspects of a jockeying for position – to offer constitutional guarantees that met Norwegian concerns led to the end of resistance.

In 1808, Napoleon moved most of his forces to northern Spain and in November marched south, defeating the poorly trained and commanded (and outnumbered) Spanish forces at Espinosa, Gamonal and Tudela, and entering Madrid. This was the only occasion on which Napoleon campaigned in Spain. However, his operational victory did not lead to strategic success. Resistance, encouraged by French exactions,[28] continued, and Britain sent forces, helping make the Peninsular War (1808–14) a long-term drain on the French army. Meanwhile, the Continental blockade Napoleon had declared on Britain in an effort to wreck her economy by exclusion from European markets and supplies was unsuccessful, and, indeed, helped to exacerbate discontent with French power. With his failure to take over Spain, a major naval and imperial power, without a major rebellion, Napoleon lost his opportunity to continue a major maritime and transoceanic struggle with Britain. The process by which other Western transoceanic empires came into the British orbit through conquest, intimidation, alliance protection or trade, or by a combination of some of these (as with the uninvited British occupation of the Portuguese colonies of Goa from 1799 to 1813 and Macao from 1808 to 1811), was greatly pushed forward.

Moreover, for France, there was no equivalent to the thirteen colonies in North America during the Anglo-French war of 1778–83 to provide France with assistance within the British world. In contrast, Irish disaffection after the rapidly suppressed rebellion of 1798 was small-scale. Nor was separatist feeling in the Portuguese and Spanish colonies in Latin America yet sufficiently developed to provide France with allies, or, at least, diversionary support.

Instead, this separatism offered a base for continued opposition to France and its takeover of Spain, although, subsequently, the Spanish Bourbons were to find it impossible to reimpose their authority for other than brief periods and by 1824, when they were defeated in Peru, had been driven from their mainland colonies.

Furthermore, although relations between Britain and the USA were increasingly strained in the late 1800s, there was no outbreak of hostilities until 1812, and the Americans did not look to French support against Britain. Good American–French relations did not follow the sale of Louisiana in 1803. Instead, that was part of a process of withdrawal from hopes of a French empire in the west which had been designed to include also Florida, Cayenne, Martinique, Guadeloupe and Saint-Domingue, where there had been a major slave rebellion in 1791. In 1802, Napoleon sent 20,000 troops to suppress opposition there, but resistance continued, the resumption of war with Britain in 1803 led to a blockade, and the French were obliged to evacuate that year, the independence of Haiti being proclaimed on 1 January 1804.

Napoleon, moreover, failed to create an alliance system among non-Western powers. In 1798, his secretary, Louis Antoine Fauvelet de Bourrienne, thought that the general wished to repeat the triumphs of Alexander the Great by marching overland against India. Failure in Palestine in 1799, and, two years later, of the force that Napoleon had left in Egypt, did not end his thoughts of the East. In 1803, General Decaen sailed from Mauritius to try to reoccupy Pondicherry, which had been France's major base in India. His squadron carried some 1,250 men, mostly young officers intended to continue France's earlier policy of raising and training forces for native rulers. They could not land at Pondicherry, however, because it was occupied by the British, and that year General Lake destroyed the 'Brigades françaises de l'Hindoustan' of Boigne and Perron in North India. A French attempt in 1806–7 to develop an alliance with Persia, which included the dispatch of a military mission designed to reorganise the Persian army, failed due to contradictory goals and was wrecked by the French alliance with Russia in 1807.

Napoleon's failure on the wider world scale was both cause and effect of British maritime superiority, although the key period in Asia was 1798–1803, when Britain successively overawed or defeated Hyderabad, Mysore and the Marathas. Like Louis XIV, Napoleon wanted colonies and a strong navy, but, under pressure, the army came first. This had strategic consequences not only in terms of alliances but also with reference to the profits to be gained by oceanic trading systems. Mauritius and Réunion were important French bases, but they did not produce the goods and silver to trade with China that the British gained from India and its trade. Egypt was seen as a source of cotton, rice and coffee, and as a base for French commercial expansion, but Napoleon was unable to retain control after his initial success in 1798. Within Europe, far from funding allies, Napoleon, instead, had to use them as a source of support, a marked contrast with the situation in the French-alliance system during the Seven Years' War. More generally, the effects of his policies increased deindustrialisation and an emphasis on war as the source of profit and prestige.

In April 1809, Napoleon confronted a rearmed Austria which, concerned about its safety in the face of French policy, proclaimed a War of German Liberation. Austrian forces invaded France's ally Bavaria, only to be driven out by an effective French response as Napoleon seized the initiative beating the Austrians at Abensberg (21 April), Eckmuhl (22 April) and Ratisbon in a series of clashes, rather than a major battle. Prussia also failed to support Austria, which was handicapped by poorly conceived war aims, inadequate and divided central leadership and a foolish strategy. Attempts to launch popular uprisings in Germany failed.

Napoleon then invaded Austria. The first major battle there, Aspern-Essling (21–22 May), reflected little credit on Napoleon, whose over-bold, inadequately reconnoitred and poorly thought-out advance on a superior Austrian force under Archduke Charles was repelled and followed by a serious Austrian assault on the French, who were isolated on the north bank of the Danube, with the main bridge smashed. The French army was not destroyed, but Napoleon had to abandon the battlefield. He counter-attacked at Wagram on 5–6 July, but, despite the fact that he drove Charles from the field, the Austrian army was not routed. At Wagram, the French flank offensive was successful, the Austrian one not, and both sides used artillery to great effect. Austrian effectiveness in these battles has been seen as a stage in the emergence of modern war,[29] which places too great a weight on one campaign, but reflects the extent to which Napoleon was facing more impressive opponents.

The nationalist risings in Tyrol and northern Germany did not affect the campaign. Hofer's rising in Tyrol, which had been transferred to Bavarian rule in 1805, had been initially successful, but it never received effective Austrian support, and Bavarian and French forces eventually reconquered the region. However, they only did so after bitter conflict, including a number of serious checks for the Bavarians, which indicated the difficulty of counter-insurgency warfare. This difficulty led most troops to prefer conventional battle, even one as deadly as Wagram.[30]

Wagram was followed by peace with Austria on French terms, the Treaty of Vienna of 14 October 1809. This led to an extension of French territory and a limitation on the size of the Austrian army. France had used her military force to overrun Germany and Italy, achieving a hitherto unprecedented power. The war had also tested the Grand Empire. Half of Napoleon's available forces in Germany used to counter the Austrian offensive were in fact German. The Rhenish princes held to their agreements with France. Furthermore, Alexander I held to the Russian agreements at Tilsit and Erfurt, committing a small army for the occupation of Galicia, the Austrian part of Poland.

Conflict, however, continued in Iberia (Spain and Portugal), with British, Portuguese and Spanish forces proving intractable opponents to France in the

Peninsular War. Wellington's victory at Talavera (1809) led Napoleon to appreciate the seriousness of the situation. The defeat of Austria was followed by the dispatch of substantial French reinforcements, and in 1810 a major attempt was launched to drive the British from Iberia. Thanks to defensive preparations, especially the Lines of Torres Vedras, Wellington's skill, and the support provided by control of the sea, which permitted the supply of British forces, the French failed. Wellington ably executed fire-and-movement tactics, balancing the well-drilled line with the extensive use of light infantry, the conservatism of an emphasis on linear fire-power formations with a greater role for manoeuvrability. He never had more than 60,000 British troops under his personal command and was always heavily outnumbered in both cavalry and artillery, but his troops were among the best in the British army. Wellington was also a fine judge of terrain and adept at controlling a battle as it developed.

The French conversely suffered from sometimes-indifferent command, as well as the frequent unwillingness of their generals to cooperate and Napoleon's inappropriate interference. The French also had inadequate battle-field tactics, relying on crude attacks in dense columns which provided easy targets for the British, as at Vimeiro (1808), Talavera and Bussaco (1810). Wellington was also very active in counter-attacks, and the well-timed bayonet charge, launched when the French were disorganised by their approach march and by British fire, was as effective as the volley. Medical records on casualties, and other sources, suggest that the bayonet was essentially a psychological weapon in most Napoleonic engagements. Firepower caused more casualties and was therefore crucial to the decision of the battle. However, the bayonet charge permitted exploitation of the advantage. Such a charge, preceded by a volley, had become a standard British tactic from the late 1750s, used with effect in the American War of Independence, and, with his fine grasp of timing and eye for terrain, Wellington brought the system to a high pitch of effectiveness. At Salamanca (1812), Wellington rapidly and effectively switched from defence to attack, and the offensive was also key to British victories at Vitoria (1813) and Toulouse (1814). Casualties, however, were often heavy: more than a quarter of the British force at Talavera and 40 per cent of those at Albuera (1811).

There was no major conflict further east until Napoleon invaded Russia in 1812, but, meanwhile, France's relative advantages had eroded. The French lacked a lead comparable to that enjoyed (although not without anxiety) at sea by the British after Horatio Nelson's shattering victory over a Franco-Spanish fleet at Trafalgar in 1805. As so frequently in Western military history, a capability gap on land had been swiftly closed (although the extent of such a gap anyway between France and Russia is questionable). The Prussians made a major effort to incorporate skirmishers into their battle formations and also developed both a reasonably flexible general staff system and an effective militia,[31] while Austria adopted the corps system, although Spanish attempts first to improve the army and, from 1808, to create a new one, were less successful.[32] Combined with the widespread reluctance within Europe to accept French aggression and expansionism, the closure of the capability gap ensured the defeat of Napoleon's unrealistic drive for hegemonic power, a drive that proved incompatible with contemporary assumptions of European order.[33] In 1813–14, Napoleon was no longer fighting the Eastern powers (Austria, Prussia, Russia) one at a time.

Tsar Alexander, who had been forced to accept French domination of most of Europe when he met Napoleon at Tilsit in 1807, was increasingly concerned about French strength and intentions, while Barclay de Tolly had reformed the Russian army. The Russians were also successful in conflicts with Turkey (1806–12) and Sweden (1808–9), emerging with the gains of Bessarabia and Finland, and with their flanks strengthened. As with his attacks on Austria, Prussia and Spain, Napoleon resolved to deal with a crisis he had played a major role in creating by invading his opponent and striking at the centre of its power. He sought to do so with an army so large that it would guarantee victory. Over 600,000 men were available for the invasion: 200,000 of them were French, another 100,000 from departments annexed to France after 1789, and the remainder were allied, principally German, forces, including 36,000 Bavarians, 34,000 Austrians, 27,000 Saxons and 20,000 Prussians. Alexander was outnumbered, but, as a result of this concentration, the flow of French reinforcements to Spain was stopped and Wellington was therefore able to launch an offensive there.

On 24–25 June 1812, Napoleon's forces crossed the Niemen river without resistance. The Russians fell back, denying Napoleon, who entered Vilnius on 28 June, the decisive battle he sought. The French, in pursuit, were faced by growing supply problems – exacerbated by Russian scorched-earth and guerrilla actions – and by the loss of men and horses through disease, fatigue, heat and hunger, which led Napoleon to send increasingly frantic demands for reinforcements. Dysentery and typhus hit the army. The Dnieper was crossed successfully, but attacks on the Russian forces defending Smolensk failed (17 August), and the Russians withdrew successfully. The Russians sought to stop the advance on Moscow in prepared positions at Borodino (7 September), a battle involving 233,000 men and 1,227 cannon. Instead of seeking to turn either of the Russian flanks, Napoleon attacked the Russian left, and the battle lasted all day. The Russians resisted successive French attacks on their entrenchments and were driven back without breaking. Napoleon refused to commit the imperial guard, which might have had a decisive impact, and the Russians abandoned the battlefield at night. Despite the heavier Russian casualties, it was Napoleon's losses, about a quarter of his army, which were crucial. Borodino was very much a battle of attrition, with total casualties amounting to some 77,000 men.

The road to the capital was now open. On 14 September 1812, Napoleon entered an undefended Moscow, but with just under 100,000 troops, and only to find the city set ablaze that night (probably by the Russians) and Alexander refusing to negotiate. In the face of a very nasty strategic situation, and with the supply situation continuing to deteriorate, Napoleon abandoned Moscow on 19 October. Heavy snowfalls from 4 November, however, turned the retreat into a nightmare as the French supply system collapsed, and exposure and starvation carried off thousands. The Russian attempt to cut Napoleon off at the Beresina river (26–29 November) failed, but the French rearguard suffered heavily there and the army that left Russia had suffered more than 300,000 casualties (including 31,400 of the 36,000 Bavarians), as well as losing about 1,000 cannon and 180,000 horses. The Russians had also lost heavily, maybe 250,000 troops.

Opposition to French hegemony required a skilful response, both military and political, but Napoleon had failed to provide this. Strategically, operationally and, at Borodino, tactically, he had mismanaged the 1812 campaign. His inability to translate the capture of Moscow into Russian acceptance of his hegemony (Alexander rejected negotiations) and his retreat from Moscow at the end of the 1812 campaign, with his army ebbing away into the snows, were fitting symbols of the folly of his attempt to dominate all of Europe.

In 1813, Napoleon wrote 'War is waged only with vigour, decision and unshaken will. One must neither grope nor hesitate.' The year began, however, with his demoralised and indifferently commanded forces retreating across Poland very much against Napoleon's will.[34] More seriously, his diplomatic position collapsed. The end of war with Turkey in 1812 helped the Russians focus on France. In March 1813, Prussia declared war and, by using its *Landwehr,* a militia which could serve as frontline troops, created a large army of 272,000 troops. The weakness of Napoleon's position had been concealed in 1809 by the success of his operations and by Austria's isolation, but in 1813 the situation was different, not least with more determined opponents.[35] Napoleon rebuilt his army to a field force of over 400,000 plus his artillery, but the new recruits were more like the fresh troops of 1792 than the veterans of his earlier campaigns. In particular, Napoleon was unable to create a new cavalry to match the troops and horses lost in Russia. More generally, as the number of French veterans was cut by casualties, the proportion of new recruits in Napoleon's army rose. As a result, the proportion of French soldiers accustomed to manoeuvring under fire decreased, and this may have contributed to the stress on numbers and straightforward assaults. His victories over the Prussians and Russians at Lützen (2 May) and Bautzen (20–21 May) were achieved over outnumbered forces, and neither was decisive; although they might well have brought victory had Napoleon pressed on. Bautzen, instead, led both sides to agree to an armistice.

Napoleon rejected peace terms that summer, but Austria (on 12 August) and Sweden joined his opponents, and Napoleon was now heavily outnumbered, by over 600,000 to 370,000 in his total field army. The Allies adopted the Trachenberg Plan: battle with Napoleon was to be avoided while independent forces under his subordinates were to be attacked, a policy Napoleon facilitated by his willingness to detach forces in order to try to capture Berlin, a political objective which contrasted with the supposed Napoleonic focus on knocking out the opposing army.[36] The Prussians defeated detached French forces at Grossbeeren (23 August), on the Katzbach river (26 August), at Hagelberg (27 August), and at Dennewitz (6 September); and the Austrians won at Kulm (30 August). Napoleon's failure to train his marshals to operate as independent commanders, and their lack of supporting staffs comparable to that which supported Napoleon, cost the French dear. The marshals also could not cooperate to fulfil strategic objectives. The scale of war was too great for Napoleon to control everything.

Only at Dresden, where Tsar Alexander and Frederick William III insisted on fighting on 26–27 August, was Napoleon victorious, thanks to strong attacks by his flank forces. Nevertheless, this was a tactical not an operational envelopment and not the triumph that the French required were they to win. By failing to concentrate his forces, Napoleon had allowed their attenuation, and this had preserved neither the territory under French control nor the strategic advantage. Instead, it was Napoleon who had lost the initiative and was outmanoeuvred, his line of retreat threatened by the converging Allied forces. Attrition was thus combined, by Napoleon's opponents, with operational advantage, while their increased military effectiveness was also important.

At Leipzig (14–19 October), the 'Battle of the Nations', the French were outgeneralled and outfought in the most serious defeat in battle that Napoleon had hitherto faced. At the outset of the battle, he exploited the advantage of the interior lines on which he was operating, while his mutually suspicious opponents faced the difficult task of cooperation on exterior lines, although this provided the opportunity of encircling the French and cutting their lines of communication. Napoleon was unable to exploit the situation in order to win. His opponents were not defeated in detail (separately). The battle posed serious problems of command and coordination for both sides, with a range of struggles waged across a considerable area: alongside large-scale assaults, there was extensive use of artillery, and also major cavalry attacks, particularly by the French on 16 October. Under pressure from greater Allied numbers, which became more of a threat as they converged, particularly once the 60,000 strong Army of the North arrived, the French were finally defeated. Unable to defeat his opponents, whom he nevertheless held off, Napoleon retreated, but the premature destruction of the Elster bridge trapped four corps. Defeat led to 73,000 casualties,[37] as well as the collapse of Napoleon's position in Germany, as former allies, especially Bavaria, Saxony and Württemberg, deserted; although Napoleon was able to thwart an attempt to block his retreat into France.

The 1813 campaign clearly indicated that symmetry had revived in Western warfare. It was possible to defeat the main French field army, and without the benefit of the distances and climate of Russia. The aura of Napoleonic invincibility had been wrecked by the Russia campaign, the Austrians had now learned to counter the French corps system by using one of their own, and the Prussians had improved their army, not least by developing a more coherent and comprehensive staff system. British successes in Spain compounded Napoleon's woes.

In early 1814, eastern France itself was invaded by Austro-Prussian forces. There was no winter break to campaigning and this lessened the French ability to mount an adequate response. As with France during the War of the Spanish Succession, there were also serious burdens as a consequence of the need to support the military from France itself rather than from occupied or allied territories. The economic consequences of war, invasion and British blockade, and the fiscal burden of the 1812–13 campaigns, were also very serious, and this hit the *levée en masse* in 1814. So also did widespread non-compliance and resistance, which reflected the unravelling of the regime.

Napoleon, nevertheless, with some success, manoeuvred with skill in order to destroy the most exposed Austro-Prussian units. Initial victories, for example at Brienne (29 January) and Champaubert (10 February), led him to reject Allied peace proposals, but numbers told and Napoleon and his subordinates suffered defeats, as at Laon (9–10 March). In place of the 80,000 opposing troops Napoleon had anticipated, there were about 200,000, and his own army was 70,000 strong, not the 120,000 men he had hoped for. Napoleon had the advantage at the Battle of Arcis-sur-Aube (20–21 March), but he was outnumbered and, after defeating another French force at La-Fère-Champenoise (25 March), the Austrians and Prussians marched on Paris, ignoring Napoleon's position on their flank. Their united army of 107,000 men faced only about 23,000 defenders, and the latter were driven back in the Paris suburbs on 30 March, leading to the opening of negotiations on 31 March. Napoleon was still in the field when Paris surrendered, and a provisional French government deposed him. With his marshals unwilling to fight on, Napoleon abdicated on 6 April and was sent to the island of Elba, which he had been given as a principality.

Returning from exile, however, on 1 March 1815, Napoleon rapidly regained power due to the unpopularity of Louis XVIII, and by 20 March he was back in Paris. Four days earlier, Austria, Britain, Prussia and Russia had already promised to keep a combined army of 600,000 men in the field until Napoleon was defeated. The overwhelming strength at their disposal led Napoleon to strike first at his nearest opponents: a British-Dutch-German army under Wellington at Brussels, and the Prussians under Blücher at Liège. On 15 June 1815, Napoleon invaded Belgium and gained the strategic advantage. The following day, his forces engaged Blücher at Ligny and Wellington at Quatre Bras. Blücher was defeated by Napoleon with heavy casualties (although this was insufficient to end Prussian fighting potential), but Marshal Ney had less success against Wellington. However, exposed by Blücher's defeat, Wellington fell back on a ridge at Mont St Jean.

The subsequent battle of Waterloo (18 June), the last of the sequence, found Wellington with 68,000 troops holding his position against attacks by Napoleon's 72,000. British defensive firepower beat off successive, poorly coordinated French frontal attacks. The French army was not at its peak, nor was it well commanded. Napoleon, who was ill, underestimated Wellington's generalship. The individual French arms were not combined ably, and there was a failure to grasp tactical control. The British line was not weakened by prior engagement. Yet, for all Napoleon's failings, the French were a formidable army, willing to take heavy casualties, and Wellington regarded Waterloo, in which he suffered over 15,000 casualties, as his hardest battle, and described it as a 'pounding match'. Flank attacks, which Napoleon neglected, or yet more frontal assaults, might have succeeded, but the arrival of plentiful Prussian support in the latter stages of the battle helped Wellington greatly and threatened the French with envelopment. Wellington still had reserves, although by the time Prussian intervention made a major difference, the situation was very difficult for him. Napoleon, however, had to deploy many of his reserves against the Prussians.[38]

After the battle, France was easily invaded. Napoleon abdicated on 22 June, Paris was occupied on 7 July and Napoleon surrendered to a blockading British warship. The Second Treaty of Paris of 20 November stipulated an occupation of northern France for five years, a large indemnity of 700 million francs and the cession of the towns of Beaumont, Bouillon, Landau and Saarlouis. By the Quadruple Alliance of 20 November 1815, the four great powers – Austria, Britain, Prussia and Russia – renewed their anti-French alliance for twenty years, a step designed to limit the chances of France disrupting the alliance. Napoleon had failed, totally. The contrast with the more successful Louis XIV is instructive,[39] while Louis XV constitutes a half-way house, his military able to achieve considerable success in Europe in 1745–8, but not to retain the French empire in 1758–62.

Napoleon has been seen both as the destroyer of eighteenth-century warfare and as its culmination. Such an exercise is inaccurate as it implies a misleading uniformity in pre-Revolutionary warfare. Furthermore, the degree of novelty can be exaggerated. Greater use of light infantry was certainly a characteristic of Napoleonic warfare, but much else that it is noted for had been anticipated in earlier conflicts: large armies, a strategy of movement, a preference for battles over sieges, and a greater emphasis on artillery. The forces involved in campaigns such as 1805, 1809, 1812 and 1813 were particularly big, but battles in the pre-Revolutionary eighteenth century already posed major problems of command and control, and many were lengthy struggles, as victory in a part of the field was countered by the moves of other formations.

It is also unhelpful to contrast Napoleonic and pre-Revolutionary warfare starkly, as conflict is a product of a multiplicity of related, but, to some degree, autonomous, activities that do not necessarily proceed on the same trajectory or with a similar pace. The Revolutionary-Napoleonic period was more of a departure in the political and social context, not least with a novel stress on propaganda, than in weaponry and naval conflict. The revolutionary ethos and purposes of the French army in the 1790s transformed the political context of military activity, freeing greater resources for warfare. The end result, however, of this warfare was to enhance British maritime and Russian land power. Both states were outside Europe, more able to protect their home base than other European countries, yet also able to play a major role in European politics. As a result, they saw off Napoleon, exploiting his inability to provide lasting stability in western and central Europe, and thus thwarted the last attempt before the age of nationalism to remodel Europe. Their success ensured the reversal of the trend towards French hegemony, which had culminated in 1810–12 with much of Europe, including the Low Countries, Hamburg, Lübeck, Savoy-Piedmont, Genoa, Tuscany, the Papal States, Trieste, Dalmatia and Catalonia, being part of France (and thus providing conscripts, firearms and supplies for her army); while client states, such as the new kingdoms of Bavaria, Italy, Saxony, Westphalia and Württemberg, were similarly created or engorged. In 1814–15, in contrast, Europe was returned by Napoleon's victorious opponents to the multiple statehood that distinguished it from so many of the other heavily populated regions of the world. This was as much a consequence of Napoleon's political failure as of the absence of a lasting military capability gap in favour of France.

10

NAVAL POWER[1]

Discussion about military change and, more specifically, concerning military revolution and modern or total warfare in the period 1775–1815, focuses on land conflict in the Western world, and generally ignores or underrates the importance of naval developments. This is unfortunate as, on the world scale, it was as naval powers that the Western states were particularly important and effective. Indeed, as throughout the period covered by this book, there was a Western naval exceptionalism that rewards attention. On land, of course, the capability of Western powers was readily apparent in the 1790s–1810s, from the Ohio to Sri Lanka and Sumatra. The newly independent Americans made important gains to the west of the Appalachians, the French conquered Egypt in 1798, the Russians defeated both Turks (1806–12) and Persians (capturing Derbent in 1796), and the British made important gains in South Asia. Each was impressive, but also needs to be placed in perspective. Aside from the defeat of attacking Western forces at the hands of non-Western powers, such as of the British in Egypt in 1807, these latter powers also campaigned actively on land against each other and with important results. These consequences included the consolidation of Vietnam in the 1790s and 1800s, the replacement of the Hausa states by the Sokoto Caliphate in West Africa, major campaigns by Burma and Siam and the development of de-facto independent Egyptian power under Mehmet Ali.

These campaigns deserve attention, not least, from the naval perspective, as the steppes of Central Asia and the sahel of West Africa can be considered as land-seas that had certain similarities to the oceans. Furthermore, although the Chinese advances against the Zunghars in the 1750s might not seem as impressive as Western transoceanic campaigns, they were formidable achievements. China under the Manchu emperors successfully solved the logistical problems central in managing steppe warfare, which was considered the supreme strategic threat by all Chinese dynasties. In the 1750s, the Chinese established two chains of magazine posts along the main roads on which they advanced. Supplies were transported for thousands of miles, while the Mongolian homelands controlled by their eastern Mongol allies provided the horses and fodder. These improvements in logistics – partly due to a desire to keep the troops from alienating the populace – ensured that the invading Chinese armies did not disintegrate as Napoleon's did in Russia in 1812, and the comparison is instructive. In order to wage war with the Zunghars, there was a massive transfer of resources from eastern to western China: the application to military purposes of the great demographic and agricultural expansion in China during the century.[2] There was a parallel with the British navy, with organisational factors in each case proving a crucial precondition for campaign success.

Non-Western navies

If Western powers shared land capability with non-Western counterparts, the situation was different at sea. It was not that the Westerners alone had naval forces. Other states also did so. These included the North African powers, Turkey, Kamehameha I of Hawaii, and a number of others, although the literature devoted to them is limited, and certainly far far less than that on Western navies. It would be mistaken to lump non-Western navies all together, as there were major differences in fighting styles and environments, as was also the case for armies. For example, the Turks were capable of fleet engagements, while, in contrast, the North African powers – Morocco, Algiers, Tunis and Tripoli, essentially deployed privateering forces appropriate for commerce raiding. The Turks and the North Africans used essentially the same maritime technology (galleys and square-riggers) as Western Europeans, although galleys became less important in the Mediterranean, and France abolished its once-important galley fleet in 1748.

Unlike in the Mediterranean, however, most non-Western naval forces were not deep-sea. Instead, particularly along the coasts of Africa, South-East Asia and the East Indies, there were polities that controlled flotillas operating in inshore, estuarine, deltaic and riverine waters.[3] These boats were shallow in draught and therefore enjoyed a local range denied Western warships. They were also quick, manoeuvrable, beachable and inexpensive. Their crews usually fought with missile weapons, which in the eighteenth century increasingly meant muskets, and some canoes also carried cannon. Similar technology was also employed in the Pacific and along the coasts of New Zealand and Pacific North America. Missile weapons could be used prior to, or in lieu of, boarding. As with land warfare in some non-Western societies, and in marked contrast to the situation in the West, the divide between conflict between humans and the hunting of animals was not too great at this level of weapon technology and military organisation.

At the same time, non-Western naval and amphibious forces were not only hunters or raiders, but also could achieve operational goals. This was seen in conflict between the New Zealand Maori in the early nineteenth century and, even more clearly, in the earlier unification of the Hawaiian archipelago. By 1789, Kamehameha I was using a swivel gun secured to a platform on the hulls of a big double canoe. Soon after, he had a large double canoe mounting two cannon and rigged like a Western schooner. Such boats helped him as he expanded his power across the archipelago. Kamehameha won dominance of the island of Hawaii in 1791, and of the islands of Maui and Oahu in 1795. In 1796 and 1809, the difficult waters between Oahu and Kauai, and outbreaks of disease, ended his plans to invade Kauai, but, in 1810, Kaumualii, the ruler of the islands of Kauai and Niihau, agreed to serve as a client king.[4]

Operational goals drew on a capacity for troop transport and amphibious operations. When James Cook visited Tahiti for the second time in 1774, its fleet was preparing for a punishment expedition against the neighbouring island of Moorea. Cook and the painter William Hodges took great interest in these war preparations, as did the general public when the painting *The War Boats of the Island of Otaheite* (Tahiti), Hodges' largest from Cook's second voyage, was exhibited at the Royal Academy in 1777. Cook estimated that the expedition involved 4,000 men.

None of this stands comparison with Western navies, but, instead of considering a single standard of capability, it is necessary to note the diversity of goals and the variety of best practice. For example, the Barbary (North African), Omani and Maratha ships were commerce raiders whose emphasis was on speed and manoeuvrability,[5] whereas the heavy, slow, big ships of the line of Western navies were designed for battle and emphasised battering power, although such a comparison is not intended to neglect the variety in the specifications and goals of these ships of the line. Furthermore, there were other types of ships in the Western navies which were not ships of the line and which were specifically designed for fast, manoeuvrable, inshore work, for example the cutters used to help customs services in the war against smugglers.

There was no revolution in Western naval capability vis-à-vis non-Western powers in this period. The strength of Western battle fleets was already apparent in the early sixteenth century, with successive Portuguese victories over Indian, Egyptian and Turkish fleets in the Indian Ocean in 1500–38, and the balance of battle advantage remained with the West, increasingly so in the eighteenth-century conflict with the Turks in the Mediterranean and, later, the Black Sea. It would be mistaken to see this advantage simply in terms of firepower. The most dramatic Western naval victory over non-Western forces in the period covered by this book – at Cesmé off Chios in 1770 – was primarily due to the effective use of fireships against the closely moored Turkish fleet by the Russians; both the method and the result were highly unusual. About 11,000 Turks were killed, although the Russians totally failed in their attempt to exploit the situation by driving the Turks from the Aegean islands.

Further afield, naval forces could also enforce Western interests. In 1725, when French merchants were expelled from their base at Mahé on the west coast of India, a squadron was sent from Pondicherry, the main French position in India. It forced the return of the merchants and obtained new commercial benefits. Moves against French trade at the coffee port of Mocha in Yemen similarly led to the dispatch of a squadron from Pondicherry in October 1736. Arriving off Mocha the following January, the French bombarded the port, disembarked troops and seized the port, thereby restoring their commercial privileges. Pondicherry, however, was to prove vulnerable to British attack in successive wars, surviving attack in the War of the Austrian Succession, but being captured in the Seven Years', French Revolutionary and Napoleonic wars.

In part because the Western powers devoted so much of their effort to fighting each other, there was no focus on naval action with non-Westerners (although there was much interest). The situation did not change until after the Napoleonic wars, when Western powers ceased their high-intensity conflicts/confrontations and, instead, there were dramatic clashes with non-Western powers including the successful British attack on Algiers (1816), the Anglo-French-Russian destruction of the Turkish/Egyptian fleet at Cape Navarino (1827), Anglo-Chinese conflict in the First Opium War (1839–42), and the Russian destruction of the Turkish fleet at Sinope (1853).

The earlier focus on conflict with other Western powers was as well, as there was no increase in relative Western capability, and thus no anticipation of the enhanced capability for inshore and riverine power projection which was to follow from the mid-nineteenth century. Such an increase, indeed, was not to occur until the application of steam power, and this bore little relationship to Western naval conflict in this period. In 1813, the American Robert Fulton drew up plans for a powerful steam-propelled frigate, significantly named *Demologos* (*Voice of the People*, launched in October 1814), but such developments still lay in the future.[6] Indeed, the early history of the steamship was a hesitant one. In 1707, Denis Papin, Professor of Mathematics at Marburg in Hesse-Cassel, demonstrated the first working steamship, the *Retort,* at Cassel, only to have the boat speedily demolished by rivermen who feared competition. It was not until 1783, on the River Saône near Lyons, that the Marquis Jouffroy d'Abbans conducted the next successful demonstration of a workable steamboat.[7]

The absence of deep-sea naval conflict not involving Western fleets in this period draws attention to Western exceptionalism and to the interacting roles of military goals, strategic cultures and interest groups. There was no inherent reasons why major non-Western powers, especially those of East Asia, should not deploy substantial fleets, and some had done so in the past, the Chinese for example in the early fifteenth century, and the Japanese and Koreans, at a shorter range, in the 1590s.[8] That they no longer did so in 1660–1815 is a subject that by its nature is difficult to probe. Aside from the conceptual problem of assessing why something did not occur, there are major difficulties in researching the subject. It is instructive, however, to consider the emphasis in recent study on Western naval developments, especially by Jan Glete and Nicholas Rodger,[9] on the cooperation of states and mercantile elites, and, in particular, on the openness of the first to advice from the latter and on an ability to derive mutual profit from naval requirements and financial resources.

Oceanic naval power depended on a maritime economy to supply resources in depth, i.e., demand from governments alone would not generate sustainable maritime infrastructure. It required both a strong state infrastructure and a strong maritime economy, and that was not only quite rare but could also be put under great pressure in war, as France discovered in conflict with Britain.[10]

In contrast to Western capability, the situation was different in the rest of Eurasia because, although Asian merchants remained important in long-distance trade,[11] the mercantile elites were generally separated from rulers by ethnic or religious divides. Linked to this, the relationship between port cities and states was often uneasy and was nothing like that represented by the role of London in British policy. Jews, Greeks and Armenians did not have close relations with the Turks, while China after the Manchu conquest in the mid-seventeenth century was very much dominated by an elite whose values were not maritime. They overcame the attempt of the Ming loyalist Zheng Chenggong to use naval strength as the basis for political power, although the potential of this course was shown in the conquest of Taiwan from the Dutch in 1661–2.[12] The same emphasis on landed values was true of the Mughals in India and of the rulers who succeeded the Safavids in Persia from the 1720s. Close attention reveals, however, that, whatever the political and governmental system outside the West, there was generally more accommodation, compromise and pressures from short-term exigencies than concern with formal structures might suggest. Provincial notables holding public office acted in an autonomous fashion. Looked at differently, this was an aspect of the flexibility of provincial administration.[13] This process certainly left plentiful 'spaces' for maritime activity.

As the newly independent USA showed, however, such maritime activity was not synonymous with naval strength, certainly of the battle-fleet type, with its focus on ship-killing, rather than the self-financing commerce raiding that it was easier to pursue in such 'spaces' and which did not require a supporting governmental infrastructure. As a consequence, attention has to return to the state level, because the decision to create such navies throws light on strategic culture, while the organisation of such forces was a formidable task. Many states lacked the requisite stability, and their government by ruler-generals accentuated the focus on land forces and campaigning, as was amply seen in eighteenth-century Burma, Persia and Siam.

China and Japan had greater governmental continuity, but this did not lead to a focus on naval forces. Japan was very much an insular state, and, if the term strategic culture means much, then it certainly pertains to the inward-looking governing elite. As a sign of widespread conservatism, most political, economic, cultural and intellectual efforts in Japan were directed at the preservation and strengthening of established arrangements, although, by the end of the eighteenth century, there was greater interest in new political, economic and cultural forms.[14]

In China, the formidable bureaucratic culture that was a legacy of the Chinese past maintained by the Manchu was not open to mercantile influences, and certainly not in comparison to western Europe. Furthermore, policy goals (themselves set in the context of socio-political assumptions and strategic culture) did not lend themselves to naval development. Having vanquished the Zunghars and occupied Kashgar in the 1750s, China did not move into a military quiescence comparable to Japan. Instead, there were a series of wars, unsuccessful against Burma in 1765–9, and Tongking (northern Vietnam) in 1788–9, and successful against Nepal in 1792. None of these required long-range naval activity: Burma was attacked overland, not by sea. With its settlement colonies on its landward frontiers,[15] China's ambitions focused on near-China, and not on distant seas. Its international relations were based on Chinese hegemony, and the offer of tribute by neighbours sufficed. Taiwan had been brought under control in 1683, there was no drive to conquer Japan, as the Mongols, once they had conquered China, had unsuccessfully sought to do in the thirteenth century, and the frontier with Russia fixed in 1689 and 1729, which excluded Russia from the Amur valley, was seen as acceptable. Neither China nor Japan challenged Russian expansion into the Aleutian Islands.

Vietnamese, Thai and Burmese actions and ambitions similarly focused on landward activities, and, despite the earlier interests of the Maratha Angria family and, to a lesser extent, the rulers of Mysore, this was also true of Indian counterparts.[16] Indian naval strength anyway was cut short by British action – against the Angrias in 1755–6 (see p. 15) and against Mysore in 1783. Persian rulers claimed hegemony over the Persian Gulf, and Nadir Shah sent a force to Oman,[17] but Persian warfare focused on conflict on landward frontiers to east, north and west – with Indian rulers, the Uzbeks, Russia and the Turks. Regional naval forces in Asia, such as those of the Buginese state of Bone in modern Indonesia, or the Illanos of the Sulu Islands, whose heavily armed galleys attacked warships of the Dutch East India Company, were only of local significance.

In so far as the development of 'modern' naval power by non-Western powers constituted a naval revolution, it needs to be dated to the late nineteenth century, although, at the turn of the seventeenth and eighteenth centuries, the Turks abandoned their traditional dependence on galleys and built a new fleet of sail-powered galleons which carried more cannon. As a result, the Turks developed an amphibious capability that was very useful for operations in the Morea in 1715. The Turkish fleet also defeated the Venetians off Chios in 1695 and checked a largely Venetian fleet off Cerigo in 1718, both important to the Turkish control of the Aegean, and each contrasting with Russian success there against the Turks in 1770. The Turks subsequently continued their borrowing, employing French experts on ship construction in the mid-1780s, although they were defeated in the Black Sea, at the battles of the Dnieper (1788) and Tendra (1790). However, no comparable response to Western naval power occurred in the Orient until the late nineteenth century.

The USA

Even in the case of the New World, where European control largely collapsed in 1775–1826, naval power in terms of significant specialised fleets did not develop until the second half of the nineteenth century. The new states focused their military activity on armies, militia ideas proved far more conducive to land than sea capability, and considerable reliance was placed on the support of European navies or unofficial support: by the Americans on France in 1778–83, and by the Latin Americans, the following century, on British support.[18] In the USA, the Department of the Navy was established in 1798, and the governing Federalists revived the navy, first to fight the Barbary States and then to engage France in the 'Quasi War' of 1798–1800. The French sank or captured over 300 American merchantmen in response to the American role in maintaining British trade routes: France did not accept that neutral ships should carry British goods. Clashes between warships from the summer of 1798 were largely won by the Americans.[19]

In the USA, the Federalist plan to build up the navy was stopped, however, when Thomas Jefferson and the Republicans gained power after the election of 1800. Jefferson favoured coastal gunboats, rather than the more expensive frigates with their oceanic range built in the 1790s. As a reminder of the political basis of strategic culture, the emphasis on gunboats conformed to the militia tradition of American republicanism, and militia could use gunboats to defend the coastal fortifications being built. Based on New Orleans, American gunboats operated against French and Spanish privateers off the Mississippi Delta in 1806–10, while others played a valuable role against Britain in the war of 1812–15.[20] Nevertheless, although, thanks to the expansion of their merchant marine, the Americans had an abundance of trained seamen to man their fleets, and the most powerful frigates of the age, which they were also adept at handling in ship-to-ship actions, they had no ships of the line, and their total navy at the outset of the war of 1812 comprised only seventeen ships. They thus lacked the capacity for fleet action. This reflected the force structure and doctrine developed under Jefferson's agrarian republicanism. It was very different in its military results to the large battle fleets created and sustained by the mercantile republics of the United Provinces and Commonwealth England in the seventeenth century.

European navies

The focus in naval power is therefore resolutely Western. Although Western warships (and merchantmen) found it difficult to operate in tropical estuary, delta and river waters,[21] there was no naval balance, nor any frontier of capability and control between Western and non-Western powers on the oceans of the world. The absence of any challenge to Western naval power on the oceans was dramatically demonstrated in the 1770s, 1780s and 1790s, as Western warships under naval commanders explored the waters and shores of the Pacific, charting and (re)naming the world, and established Western trading bases and colonies in Australasia and along the west coast of North America (the latter process had started earlier).[22] There was still much of the world's land surface where Western military strength and models were unknown, but the warships that showed their flags and ran out their guns around the globe were the forceful edge of a Western integration of the world. This, however, was not new, and, instead, had begun in the late fifteenth century.[23]

If developments within the period covered by this book do not constitute a revolution, there was still an important build-up of naval strength and expansion of capability within the existing technological constraints. There were, for example, major Western naval races in the second half of the seventeenth century (see pp. 54–5) and the 1780s. In the latter case, Britain, France and Spain all launched a formidable amount of tonnage, Spain devoting over 20 per cent of governmental expenditure that decade to the navy.[24] These huge naval forces dwarfed those of non-Western powers far more decisively than they had when Christopher Columbus and Vasco da Gama sailed forth in the 1490s. Some other powers also greatly expanded their navies in the 1780s. This was particularly true of Russia and the Dutch, which became the fourth and fifth largest naval powers. Denmark, Sweden, Naples, Portugal and the Turks also all increased the size of their navies. The total displacement of European navies rose from 750,000 tons in 1770, to 1 million tons by 1760 and 1.7 million tons in 1790.[25]

Alongside new shipping, there was the continual effort required to maintain and repair ships, an effort that in part stemmed from the vulnerability of ships' organic properties, especially wood and canvas. As the specifications of warships changed relatively little, ships could be kept in the line for decades as long as they were kept seaworthy.

Programmes of naval construction and enhancement registered not only the growing resources of Western governments but also the capability of their military-industrial complexes and the ability of their administrative systems to plan and implement changes. Fleets of warships were powerful and sophisticated military systems, sustained by mighty industrial and logistical resources based in dockyards that were among the largest industrial plants, employers of labour and groups of buildings in the world, for example Portsmouth, Plymouth, Brest, Toulon, Ferrol, Cadiz and Karlscrona, the last the third most populous town in Sweden in 1700.[26] These dockyards were supported by massive storehouses, such as the vast Lands Zeemagazijn in Amsterdam, which was destroyed by fire in 1791. The 1,095-foot-long ropery opened at Portsmouth in 1776 may well have been the largest building in the world at the time. Naval bases also required considerable investment. Developments in government policy and strategic intent led to new facilities. The substantial expansion of Portsmouth dockyard in the 1690s (the addition of two new dry docks and two wet docks) and the creation of a new front-line operational yard at Plymouth permitted the stronger projection of English naval power into the English Channel and the Western approaches in the 1690s.[27]

Similarly, the major expansion of Russian naval power under Peter the Great was linked to the foundation of St Petersburg as capital, 'window to the west', and port, on Russia's newly conquered Baltic coastline. In 1703, Peter himself laid the foundation stone of the Peter-Paul Fortress. The following year, he founded the Admiralty shipyard on the bank of the River Neva opposite the fortress, and in 1706 its first warship was launched. A naval academy followed in 1715. Furthermore, as soon as the Russians had seized a Black Sea coastline and the Crimea in 1783, they began to develop bases there, particularly at Kherson, Sevastopol and Odessa. These threatened a direct attack on the Turkish capital, Constantinople, providing the Russians with a strategy different to that of a land advance across the eastern Balkans. As a result, Sevastopol was to play a key role in Allied goals during the Crimean War (1854–6). Catherine II took the visiting Joseph II to visit Sevastopol in 1787.

The establishment of military-industrial complexes also demonstrated a more widespread capacity to stimulate change. Improved infrastructure and better naval construction lessened the problems of decay caused by the use of what were organic working parts.[28] There were also numerous innovations, which were put to good use. As an instance of the importance of incremental improvements, British cannon fire proved particularly effective in the victory over the French off the Îles des Saintes on 12 April 1782. These improvements, including in flintlocks, tin tubes, flannel cartridges, wedges to absorb recoil and steel compression springs, increased the ease of serving cannon, of firing them instantaneously and the possible angles of training.

Moreover, the location of the battle, south of Guadeloupe in the West Indies, was a testimony to the importance of transoceanic operations: in 1759, in contrast, the key naval battles of the Seven Years' War, Lagos (off Portugal) and Quiberon Bay, had been fought in European waters, as the central issue was the threat of a French invasion to Britain, which appeared to be the sole way to counter Britain's transoceanic operations as well as its key role in supporting opposition to French interests in Germany.[29] Similarly, the key naval battles in the Dutch, Nine Years', Spanish Succession, Quadruple Alliance and Austrian Succession wars had all been in European waters. Beachy Head (1690) and Barfleur (1692) arose over the threat of French invasion of Britain, while Malaga (1704), Cape Passaro (1718), Toulon (1744) and Minorca (1756) reflected the attempt to use naval force in order to affect Mediterranean amphibious operations, and the two battles off Cape Finisterre (both 1747) were related to commerce interception. The last two demonstrated that the French fleet could no longer escort major convoys bound for French colonies, and this destroyed the logic of the French imperial system. Battle alone was not the key: strategic advantage without battle was demonstrated in 1745–6 when the British navy joined the weather in dissuading the French from invading Britain in support of the Jacobites. In response to the latter, the British navy also covered the movement of troops back from the Low Countries in 1745 and the supply of advancing forces in eastern Scotland in 1746. On other occasions, these functions led to naval battles, as in 1716 at Dynekilen, where the Danish fleet defeated a Swedish supply fleet, ending Charles XII's attempt to capture Fredriksen that year. Powers that were at peace could also use naval demonstrations in order to further their views, as when France sent squadrons to the Baltic in 1739 and the Caribbean in 1740.

During the eighteenth century, improvements in seaworthiness, stemming in part from the abandonment of earlier top-heavy and clumsy designs, increased the capability of warships, both to take part in all-weather blockades and to operate across the oceans. The emphasis on maximising firepower in the late seventeenth century (see p. 32) led to a development of three-decker capability. In the early eighteenth century, the focus instead was on stability, range and versatility, which led to a move towards two-deckers, but three-deckers became important anew for the bruising confrontations of the late eighteenth century. In the case of Britain, George, Lord Anson, First Lord of the Admiralty in 1751–62, and the outstanding designer Sir Thomas Slade, were responsible for valuable innovations in warship design. The old ship types, with eighty, seventy and sixty guns, were abandoned in favour of seventy-four and sixty-four gunners (fourteen of the former were in service by 1759); the fifty-gun ship was discarded as a ship of the line, but retained in limited numbers as a heavy cruiser; the small two-deck cruisers of forty-four guns were abandoned, in favour of the single-decked thirty-two-gun frigate, and better three-deckers were also designed.

These changes further helped increase the tactical and operational effectiveness of the British navy. Earlier, greater operational capability had encouraged the projection of large-scale naval power. A large English fleet was dispatched to the Mediterranean in 1694 and wintered at Cadiz. In both 1694 and 1695, the fleet stiffened the Spanish resistance to French pressure on Catalonia.[30] The British fleet also repeatedly played a major role in Mediterranean power politics in 1704–20, and British neutrality in the War of the Polish Succession (1733–5) helped explain the greater success of Spanish operations in Italy compared to the situation during the War of the Quadruple Alliance (1718–20). As an island, Britain had little choice but to concentrate on maritime power, whereas countries with significant land borders could look inland for trade and conquest.

During the American War of Independence (1775–83), the British navy responded to the crisis both by arranging a major programme of construction and by technological advances. Copper sheathing reduced the difficulties caused by barnacles, weeds and the teredo worm, and the consequent loss of speed, and made refits easier. It was pressed forward from February 1779 by Sir Charles Middleton, Comptroller of the Navy from 1778 to 1790. In 1780, forty-two ships of the line were given copper sheathing. Politicians noted a sense of new potential, Charles, 2nd Marquess of Rockingham claiming in

1781, 'The copper bottoms occasioning our ships to sail so much better enables us either to go and attack if we should see an inferior fleet or to decline the attempt if we should see a superior fleet.'[31] The value of copper sheathing can be questioned, but the administrative achievement was considerable. The introduction of the carronade, a new, light, short-barrelled gun that was very effective at close quarters,[32] adopted by Britain in 1779, was also important: it was used with effect at the Saintes in 1782. After the American War of Independence, the French adopted recent British naval innovations, such as copper-sheathing. Standardisation, furthermore, was increasingly apparent prior to the changes brought in the late nineteenth century by the universal adoption of steam power and armour plating: in 1786 the French adopted standard ships designs for their fleet.

More generally, progress in British metallurgy improved their gunnery towards the end of the century. Britain had an advantage in technology in the shape of more powerful guns, as well as benefiting from superior seamanship and well-drilled gun crews. This had a major effect on effectiveness. The impact of British naval gunfire on enemy hulls and crews markedly increased during the war period 1793–1815, with enemy ships being reduced to wrecks in a comparatively short time.[33] The need for cannon helped drive the growth of the iron industry.

The incremental process of naval improvement continued in the last decades of sail, but, nevertheless, with the hindsight provided by consideration of subsequent technological developments, it is possible to see the period in terms of the use of yet greater resources of people, matériel and funds to pursue familiar military courses, a comment also true for much of the warfare on land. The American and French Revolutions, however, did not bring changes in naval warfare comparable to those on land. Instead, the long-term growing stress on naval firepower continued to affect fleet structures. Whereas in 1720, there were only two warships displacing more than 3,000 tons, by 1815 nearly a fifth of the naval strength above 500 tons was in this category. In 1800–15, ships of 2,500–3,000 tons achieved greater importance, whereas those of 2,000–2,500 and 1,500–2,000 tons declined in number. These bigger ships were able to carry heavier guns. Whereas the average ship of the line in 1720 had sixty guns, and was armed with 12- and 24-pounders, that of 1815 had seventy-four guns, with 32- and 36-pounders on the lower deck. Nevertheless, this greater firepower, which did not begin in the Revolutionary-Napoleonic period, did not lead to dramatic changes in naval warfare.[34]

There were other improvements. Better signalling in the period 1790s–1810s helped to enhance the potential for tactical control. The invention of a system of ship construction using diagonal bracing in order to strengthen hulls and to prevent the arching of keels, was to increase the resilience of ships, and thus their sea- and battle-worthiness, and to permit the building of longer two-deckers armed with eighty or ninety guns. These improvements helped make earlier ships appear redundant, certainly for the line of battle, but, although Robert Seppings, Surveyor of the Navy from 1813 to 1832, experimented in the 1800s at Plymouth and Chatham, the first ship built entirely on this principle, HMS *Howe,* was not launched until 1815. Diagonal framing was mainly significant after the introduction of steam made it important to build longer ships. More mundanely, but also as part of a general process of improvement, there were earlier developments in fittings, for example new patterns of anchors and the first chain cable, as well as iron watertanks in place of wooden casks.[35]

Alongside incremental improvements, progress did not constitute a steady chronological continuum. The entire business of being at sea and fighting at sea fluctuated for a number of reasons, not least of which was the impact of peace, when most of the lessons learnt in the last war were usually forgotten, albeit briefly. Furthermore, the lack of a structured and rigorous system of officer training in the skills of command led to a wide variety in command competence and method throughout the period: the situation was still bad for the British at Trafalgar in 1805. A study on how things were done reveals many ups and downs within a slow trajectory of improvement.[36]

Westerners also took their naval military-industrial capability abroad, developing major shipyards at colonial bases, such as Bombay and Halifax for the British, and Havana, where the Spaniards made good use of tropical hardwoods, producing particularly good ships, including several of their larger warships which fought at Trafalgar. These bases were important in helping to make empires systems of power and also played a key role in local economies. Halifax naval yard, founded in 1758, was the largest industrial centre in British North America after the American War of Independence.[37] In the West Indies, the British had two naval bases on Jamaica – Port Royal and Port Antonio – as well as English Harbour on Antigua, begun in 1728. Port Royal was able to careen the larger ships of the line sent there. The growing British naval and mercantile presence in the Indian Ocean owed much to shipyards in India, where merchantmen averaging 600–800 tonnes and capable of carrying very large cargoes were constructed, as well as naval vessels, including several ships of the line. Batavia (Djakarta) was the key naval base for the Dutch in the region, and Port Louis on Mauritius for the French.

The limited availability of bases and, more generally, logistical limitations, however, along with disease and climate, substantially circumscribed Western power projection outside European waters. Notwithstanding the use of Halifax, Bermuda and Jamaica, the British navy lacked the requisite support bases to mount an effective blockade of the east coast of North America, either in 1775–83 or in 1812–15, although it could inflict considerable damage on the American economy.[38] Indeed the course of the American War of Independence indicated the limitations of naval and amphibious power. The campaigns that led to the British relief of Quebec in 1776 and their capture of New York (1776), Philadelphia (1777), Savannah (1778) and Charleston (1780), and to the Franco-American concentration against the British in Yorktown in 1781, each reflected, at least in part, the amphibious capability stemming from naval strength and drawing on a development of relevant doctrine.[39] Nevertheless, in each case, the exploitation of this capability was dependent on the campaigning on land. More generally, British naval strength could not ensure a decisive victory over the main American field army.[40]

This factor, and the consequences of French entry into the war in 1778, were important to the course of the war; whereas American experimentation with new naval technology in the shape of the submarine had no effect. The first known description of a viable submarine was published by the English mathematician William Bourne in 1578, but an American, David Bushnell (1740–1824), constructed the first operational machine. In 1774, he began to experiment with a submersible that would plant gunpowder beneath a ship; the following year, a prototype was ready for testing and a way had been found to detonate a charge underwater. The wooden submarine, in effect an underwater minelayer, contained a tractor screw operated by hand and pedals, a surfacing screw, a drill for securing the explosive charge (fitted with a time fuse) to the hull of the target, a depth pressure gauge, a rudder with a control bar, bellows with tubes for ventilation, ballast water tanks, fixed lead ballast, detachable ballast for rapid surfacing, and a sounding line. The same year, Joseph Belton of Groton presented to the Pennsylvania Committee of Safety his plan for a submersible, which was expected to hole warships below their water line. The boat was designed, unlike Bushnell's, to carry one or more cannon.

These ideas were genuinely revolutionary and offered a mode of warfare against which the British had no defence. Bushnell's *Turtle* could only attack ships at anchor, but, even so, the anchorages of the British fleet provided obvious targets. However, successful execution was a different matter. The *Turtle* was first employed against HMS *Eagle* in New York harbour on 6 September 1776, but Bushnell encountered serious problems with navigating in the face of the currents and could not attach the charge, which went off harmlessly in the water. The second attempt, against HMS *Phoenix* on 5 October 1776, also failed: the *Turtle* was spotted, Bushnell's depth measurer failed, and he lost his target. George Washington pointed out the difficulty of operating the machine satisfactorily, and it is not surprising that Bushnell received scant support from the hard-pressed government, while Belton had no success.

French entry into the war in 1778, followed by that of Spain in 1779 and the Dutch in 1780, totally altered the maritime situation, leading to a worldwide naval conflict, at once more extensive and in some spheres, such as the Bay of Bengal, more intensive than previous maritime wars between Western powers. British warships were redeployed, as Britain responded to the integration of the American conflict into a wider struggle, in which the naval balance in American waters was interrelated with that in European and, more obviously, Caribbean waters. By 1780, thanks to shipbuilding since 1763, France and Spain had a combined quantitative superiority over Britain of about 25 per cent, and, partly as a result, Britain gained control of neither

European nor American waters, and was unable to repeat its success in the Seven Years' War of 1756–63.

The American War of Independence posed serious problems of naval strategy for Britain, France and Spain, although, in response, there was no revolution in strategy. For Britain, the problem of numbers of warships interacted with disputes over strategy, as the desirability of blockading French ports, for which there were arguably too few British ships, clashed with the prudent argument of John, 4th Earl of Sandwich, the First Lord of the Admiralty, that naval strength should be concentrated in home waters, not only to deter invasion but also to permit a serious challenge to the main French fleet, which was based nearby at Brest, and thus to gain a position of naval dominance. This goal would be compromised by dispersing much of the fleet among distant stations, where it could support amphibious operations and protect trade but could not materially affect the struggle for naval dominance. Due to the state of communications technology, a situation which was not to change radically until the use of radio in the early twentieth century, the commanders of those distant stations were difficult to control effectively, and they jealously guarded their autonomy and resources, producing an inflexibility that was ill suited to the need to react to French initiatives.[41] The concentration of naval strength in home waters, however, ensured that France's Mediterranean naval base, Toulon, was not blockaded, and in 1778 the Toulon fleet was able to sail to American waters and threaten the British position in New York: the arrival of the French warships was the first warning that the British garrison had of the outbreak of war, but the French failed to press home their advantage of surprise.

Despite grave strategic and organisational problems, the French were more successful at sea than in the Seven Years' War, in part thanks to a determined and effective leadership.[42] Indeed, the role of the latter, a factor which is difficult to discuss or categorise in terms of the debate over military revolutions, emerges clearly in the French war effort. There is a ready contrast between the able and energetic Admiral Pierre André Suffren, who proved a persistent, redoubtable and brave opponent to the British in the Bay of Bengal and off Sri Lanka in 1782–3, and Admiral Louis-Guillouet, Comte d'Orvilliers, the commander of the attempt to invade England in 1779. This attempt was thwarted by disease and poor organisation, rather than British naval action, for, despite the hopes of George III, the outnumbered British Western Squadron under Admiral Charles Hardy failed to mount an effective response.

The battles of the period covered by the book also indicated the difficulty of achieving a sweeping naval victory, which, in successive wars, the British were not to gain until 1747, 1759, 1782, 1798 and 1805. Lacking, by modern standards, deep keels, sailing vessels suffered from limited seaworthiness, while the operational problems of working sailing ships for combat were very different from those that steam-powered vessels were to encounter. The optimal conditions for sailing ships were to come from windward in a force

4–6 wind across a sea which was relatively flat; it was more difficult to range guns in a swell. Limitations on manoeuvrability ensured that ships were deployed in line in order to maximise their firepower, and skill in handling ships in line or in battle entailed balancing the wind between the sails of the three masts in order to achieve control over manoeuvrability and speed.[43] Line tactics and fighting instructions were designed to encourage an organisational cohesion that permitted more effective firepower, mutual support and flexibility in the uncertainty of battle (see p. 32).

Tactical practice, however, conformed to theory even less at sea[44] than on land, due in part to the impact of weather and wind on manoeuvrability. The nature of conflict at sea, not least the unwieldy nature of the line, made it difficult to maintain cohesion once ships became closely engaged, and, although experience, standardisation and design improvements enhanced performance, there were still significant limitations.[45] The report on the indecisive battle with the French Brest fleet off Ushant on 27 July 1778 by John Blankett, a 4th Lieutenant on Admiral Augustus Keppel's flagship HMS *Victory*, indicated the dependence on wind direction:[46]

> Your Lordship will recollect that the forcing a fleet to action, equal in force, and with the advantage of the wind must always be done with great risk, and our fleet was not equal to that manoeuvre, but chance, which determines many events, put it out of the Admiral's power to choose his disposition ... The French behaved more like seamen, and more officerlike than was imagined they would do ... the truth is, unless two fleets of equal force are equally determined for battle, whoever attacks must do it with infinite risk, but a fleet to leeward attacking one to windward is a dangerous manoeuvre indeed.[47]

Three years later, Admiral Thomas Graves failed to defeat the French off the Virginia Capes, an indecisive battle in terms of the damage inflicted, but, as it prevented British relief of Cornwallis's besieged army at Yorktown, an important success for the French. The engagement lasted for just over two hours, neither side having any ships sunk, but both suffering considerable damage.

Britain and France came close to renewed conflict in the Dutch Crisis of 1787 (see p. 115) and, this time with Spain on France's side, in the Nootka Sound Crisis of 1790. They did not fight again, however, until 1793, by which time the leadership and administration of the French fleet had been badly affected by the political and administrative disruption stemming from the French Revolution,[48] more so indeed than the army. In 1793, the British were invited into Toulon by French royalists, before being driven out again by Revolutionary forces benefiting from the well-sited cannon of Napoleon, then a young artillery officer.

The British enjoyed far more sweeping naval victories in fleet actions in the 1790s and 1800s than in the American War of Independence, particularly the Glorious First of June (1794), St Vincent (1797), the Nile (1798) and Trafalgar (1805). This success reflected fighting ability within a defined military system, rather than a quantum leap forward, whether described as revolutionary or not. Well-drilled gun crews, superior seamanship, bold leadership and effective command were key. Indeed, George III in 1797 had 'confidence in naval skill and British valour to supply want of numbers. I am too true an Englishman to have ever adopted the more modern and ignoble mode of expecting equal numbers on all occasion.'[49]

Thanks to its naval strength, Britain was also able during the French Revolutionary and Napoleonic wars to maintain an effective convoy system which helped them increase their proportion of world mercantile shipping and also to deny access to world markets to the French and their allies. More generally, the strength and nature of British naval power and maritime resources enabled Britain to resist Napoleon's attempt to isolate her commercially from the Continent from 1806. Naval strength also supported the British campaigns in Portugal and Spain,[50] and permitted amphibious operations, leading to the capture of French and allied overseas bases, which further lessened their ability to challenge the British: Cape Town fell in 1806, Martinique in 1809, Réunion and Mauritius in 1810 and Batavia in 1811. In 1808, Napoleon planned to take over the Spanish overseas empire, not only in the New World but also the Philippines. These hopes were thwarted by Spanish resistance, but would, anyway, have been inhibited by British naval power, just as the French attempt to regain Saint-Dominique (Haiti) from its rebellious black population had been in 1803.

Naval operations outside Europe, especially in the Indian Ocean and the Caribbean, but not only there, as the French expedition to North American waters in 1746 showed,[51] remained greatly conditioned by climate and disease. Despite improvements in some spheres, in the British case by the Sick and Hurt Board, the general conditions of service at sea remained bleak. Aside from cramped living conditions and poor sanitation, food supplies could be inadequate and inappropriate, with a lack of fresh food, fruit and vegetables, and thus of vitamin C. The cumulative impact was both to make naval service unattractive and to ensure losses among those already in service. Nevertheless, the administration was good enough to ensure that warships fulfilled their operational role on distant stations.[52]

British naval strength was also crucial in the war of 1812–15 with the USA. At sea, the British suffered initially from overconfidence, inaccurate gunnery and ships which were simply less powerful and less well prepared than those of their opponents. However, aside from three frigates lost in 1812, the British losses were all of smaller vessels, and British naval effectiveness improved during the war, both in Atlantic waters and on the Great Lakes. To operate in North America, the British were dependent both on routes across the Atlantic and on an ability to act in coastal waters. Naval blockade, which became effective from 1813, hit the American economy, amphibious forces were able to approach Baltimore and burn Washington, and it was possible to send reinforcements to Canada in order to resist successfully poorly led and uncoordinated American attacks. The American plan in May 1813 and May 1814 for a small squadron to cruise off Nova Scotia and the St Lawrence to intercept British supply ships failed in large part due to the British blockade.[53]

British naval power rested on a sophisticated and well-financed administrative structure, a large fleet drawing on the manpower of a substantial mercantile marine and fishing fleet (although there were never enough sailors), and an ability to win engagements that reflected widely diffused qualities of seamanship and gunnery, a skilled and determined corps of captains and able leadership. This was true not only of command at sea, as with Nelson's innovative tactics, but also of effective institutional leadership which developed organisational efficiency. Britain had a more meritocratic promotion system and more unified naval tradition than those of France; as well as a greater commitment of national resources to naval rather than land warfare, a political choice that reflected the major role of trade and also the national self-image. The French, in contrast, lacked an effective chain of naval command, and trade was less important to their government and their political culture than that of Britain. The same was also true of Spain and Russia.

In turn, British capability rested on the unique Western experience of creating a global network of empire and trade, which was based on a distinctive type of interaction between economy, technology and state formation, and on the specific strength of the liberal political systems, which were notably successful in eliciting the cooperation of their own and, also, other capitalists, producing a symbiosis of government and the private sector which proved both effective and, especially, valuable for developing naval strength. China, Korea and Japan could build large ships and manufacture guns, their states were relatively centralised, and their economies and levels of culture were not clearly weaker than those of contemporary Western states. However, in contrast to Western powers, there was hardly any interaction in these three states between economy, technology and state formation, aimed at creating maritime effectiveness.

The potential of Western naval power was still far less than it was to be within 105 years of the close of the period. First submarines and, subsequently, aircraft carriers led to a transformation in the relationship between warships and the environment in which they operated. The latter changed the way in which naval forces could press on land powers, as well as, more specifically, their ability to mount successful amphibious assaults. This was to be continued with submarine-launched cruise missiles. In contrast, naval power in the period under consideration was not able to have comparable impact. This was made clear by the course of the French Revolutionary and Napoleonic wars. Successive naval victories protected Britain from invasion and enabled her to risk amphibious operations, but neither these victories, nor indeed such operations, even if successful, could determine the course of conflict on land.

After Trafalgar, in which nineteen French and Spanish ships of the line were captured or destroyed, the British enjoyed a clear superiority in ships of the line, but Napoleon's victories over Austria, Prussia and Russia in 1805–7 ensured that the War of the Third Coalition ended with France in a far stronger position in Europe than at the start of the conflict. The British could not overthrow Napoleon without the help of powerful land allies. This repeated the British failure to coerce Russia into returning conquests from Sweden and the Turks, in 1720–1 and 1791 respectively.

Victory in battle, nevertheless, certainly had an operational and a strategic impact as far as the war at sea was concerned, because it greatly altered the balance of naval power. The first British naval victory of the French Revolutionary war, the Glorious First of June in 1794, helped deter France from plans drawn up a year earlier to build up a major fleet, reduced the French naval threat in British home waters, and therefore helped free more of the British navy for operations further afield, particularly in both the Mediterranean and the Caribbean. In the battle, Richard, Earl Howe, with twenty-five ships of the line, successfully attacked a French fleet of twenty-six of the line sent to escort a grain convoy from America. Howe, who had gained the weather gauge through skilful seamanship, could not fully execute his plan for all his ships to cut the French line, each passing under the stern of a French ship and engaging it from leeward. Nevertheless, enough ships succeeded, and superior British gunnery was at close range for long enough, to cost the French seven warships (one sunk, six captured) and 5,000 casualties; although the vital convoy reached France. The fate of battle partly explains the contrast between the first two years of this war, in which France lost twenty-two ships, and the first four years in the Anglo-French stage of the American War of Independence in which France only lost four warships.[54]

In 1798, British victory in the Battle of the Nile, followed by the British capture of Seringapatam in Mysore in 1799 and the killing of Tipu Sultan in his capital, and by Britain's victory over the French army in Egypt in 1801,[55] made it clear that France would not be able to project her power successfully along the Egypt–India axis. War with Britain also prevented France from enjoying the benefits of the European hegemony she seized. The British had grasped the controlling maritime position, only to see it collapse in 1795–6, when the French forced the Dutch and Spain into alliance and gained the benefit of their fleets, leading the British to evacuate the Mediterranean. Thanks, however, to British naval victories in 1797–1805, the colonial empires of France's European allies were outside her (and their) control, and the resources that Napoleon deployed could not be used to project French power overseas. This was a failure which was not inherent in France's position, but one that reflected the relatively low priority of maritime as opposed to continental activities, and the successes of the British navy.

Yet, British victory at sea could not prevent the French from trying to build up their navy, not least after Trafalgar. By 1809, the Toulon fleet was nearly as large as the British blockaders. French naval strength, however, had been badly battered by losses of sailors in successive defeats, and in 1808 Napoleon also lost the support of the Spanish navy. Nevertheless, Britain's military capability was effective in the crucial period of Napoleonic decline, 1812–14, only as part of an international league in which the major blows against France were struck on land, and by Britain's allies. Furthermore, some British amphibious expeditions, such as those against Buenos Aires and Egypt in 1807, were eventually unsuccessful, in the former case leading to the surrender of the British force. The same year, the Turks refused to yield to naval intimidation when a fleet under Vice Admiral Sir John Duckworth forced the Dardanelles, destroying a squadron of Turkish frigates. French officers then helped prepare Turkish resistance, and Duckworth deemed it appropriate to withdraw. His return through the Dardanelles proved far more difficult than his initial passage. Amphibious operations in Western waters could also be unsuccessful as with the Walcheren expedition of 1809, a mismanaged and disease-hit British attempt to seize control of the port of Antwerp and thus hinder Napoleon's attempt to rebuild his navy after Trafalgar.

Naval battles between Britain and France and her allies – Spain, the Dutch and the Danes – tend to dominate attention, but the strength and weaknesses of naval power also emerge clearly from a consideration of other aspects of naval activity. Blockade, a combination of naval strength, economic intelligence, financial pressure and diplomatic negotiation, was particularly important, with British squadrons policing the seas of Europe and, to a lesser extent, the oceans of the world. An ability to wreck the foreign trade of rivals could cripple their imperial system and greatly hamper their economy, as happened to the French at Britain's hands in 1747–8. Living standards in Danish-ruled Norway were hit hard by the blockade in 1807–14. Even if it was not possible to inflict this degree of damage, higher insurance premiums, danger money for sailors and the need to resort to convoys and other defensive measures could push up the cost of trade. George III wrote in 1795 'of the necessity of keeping constantly detached squadrons to keep the Channel, the Bay of Biscay, and the North Sea clear of the enemy's ships; had that measure been uniformly adopted by the Admiralty I am certain by this time the trade of France would have been totally annihilated.'[56]

There were different types of blockade. Close blockade was designed to stop an enemy naval force emerging, while open blockade was intended to catch an enemy naval force emerging, and maritime blockade to stop maritime commerce and to have a direct economic impact on the opponent's society, an aspect that can be related to notions of total war. These different forms of blockade had variable success rates. The history of British blockading squadrons, however, was often that of storms and of disappointed hopes of engaging the French. Blockading squadrons could be driven off station by wind and weather,[57] and blockading Toulon was particularly difficult.[58] The small watching squadron off Toulon was blown off station when the French sailed for Egypt in May 1798. The exposure of British warships to the constant battering of wind and wave also placed a major strain on an increasingly ageing fleet. The Channel fleet, for example, was dispersed by a strong gale on 3 January 1804, and the blockade of the French port of Le Havre lifted, although that of Brest was swiftly resumed. The weather claimed and damaged more British ships than the French: out of the 317 warships lost in 1803–15, 223 were wrecked or foundered, including, in December 1811, HMS *St George*, a ninety-eight-gunner, with the loss of all bar twelve of the crew of 850, and HMS *Defence*, when they were driven onto the Danish coast in a storm. Tropical stations could be particularly dangerous, and in 1807 Admiral Trowbridge and HMS *Blenheim* disappeared in a storm off Madagascar.

Fog was also a problem, particularly for blockaders. It could cover French movements, as when the Brest fleet sailed in April 1798, and, once a fleet had sailed, it was impossible to know where it had gone: in this case, the British were unsure whether the French would head for Ireland or for the Mediterranean. In January 1808, the French Rochefort squadron evaded the British blockaders in bad weather and poor visibility and sailed to Toulon, making a concentration of French warships there more serious. Fog was also a hazard to British warships. HMS *Venerable*, part of the squadron covering Brest, sank on the Devon coast in 1804 after running ashore in a thick fog. The poorly charted nature of inshore waters was a problem that led to ships frequently running aground, nearly 400 men drowning in March 1801 when HMS *Invincible* ran aground near Great Yarmouth. It was particularly easy to do so when enforcing blockades, and shoals were also a problem when attacking enemy warships sheltering in coastal waters. Once aground, ships were vulnerable to attack and weather.

Wind-powered warships were dependent both tactically and operationally on the weather. Ships could only sail up to a certain angle to the wind. Too much or insufficient wind were serious problems. Reliance on the wind alone made inshore naval operations very chancy. French ships could only leave their major Atlantic port – Brest – with an easterly wind. Due to the prevailing westerly and south-westerly winds, this was not all that prevalent, but that also created difficulties for the British blockaders who were continually blown towards the treacherous coast of Brittany, which made their lives a misery and very dangerous.

There were also serious limitations in the surveillance, and command and control capabilities of naval power. These made it very difficult to 'see' or control in any strategic sense, and certainly limited the value of any blockade. It was generally possible for a lookout to see only about 15 miles from the top of the main mast in fine weather. However, fleets used a series of frigates stationed just over the horizon, and they signalled using their sails, which were much bigger than flags, and, because the masts were so tall, could be seen at some distance over the horizon. This relay system was particularly important for blockading British fleets: there would be an inshore squadron of highly manoeuvrable ships (which were unlikely to get caught against the lee shore) which physically watched the French in Brest and Toulon, and they then signalled using a relay of frigates to the main fleet which was located a few miles off in greater safety. Surveillance capability was surprisingly sophisticated: by simply 'looking' at a ship, its nationality, strength, skill, manpower, capability and performance could all be determined.

More generally, operational limitations were tested by skill and developments. For example, specialised sailing ships, in particular bomb ketches, were designed with coastal operations in shallow waters foremost in mind.[59] It is also possible to adduce examples of successful campaigns in precisely such waters, for example the Chesapeake campaign of 1814.

Furthermore, despite its difficulties, British naval power permitted a great increase in maritime strength, and this was important to the global protection and expansion of her commerce. For example, British commercial penetration of South and South-East Asia and the Far East was aided by naval strength: occupation, as of Java in 1811–16, was important. More generally, her dominant maritime position served to ensure that Britain took the leading role in exploration, trade and the assembling of knowledge about the world. This left its mark on the imperial capital, where there was a major expansion in shipping and docks: the London Dock was excavated in 1801, followed by the West India Docks in 1802, the East India Docks in 1805, and the start of work on the Surrey Commercial Docks in 1807, all important developments in the commercial infrastructure of the empire. Elsewhere in Britain, the war led to an expansion of shipbuilding, both for the navy and for trade.

Naval power was a condition as well as a product of economic growth. As the Industrial Revolution was crucial to British and, subsequently, global modernisation, so the ability of the British navy to operate effectively, within existing constraints, in order to foster British trade was central to what was a non-military revolution. This industrial revolution was to have fundamental implications for the ability, in the nineteenth century, to develop and sustain new-model navies with totally different tactical, operational and, eventually, strategic capabilities to those hitherto.

This provides a way to consider the issues of military revolution, modernity, modernisation and total warfare that play a major role in the literature. If a technology-driven account of warfare is employed, then the naval capability of the period certainly does not deserve discussion in terms of naval revolution. Indeed, the case of the submarine demonstrated that an awareness of the difficulties posed by introducing new developments helped ensure a reluctance in adopting them. An American, Robert Fulton, produced one in 1797, but found neither France nor Britain greatly interested in its acquisition. His experiments for the French in 1800–01 included the testing of a system of compressed air in a portable container and the successful destruction of a vessel by an underwater explosion. Fulton also proposed the use of steamships for an invasion of England, but, in 1803, the French Academy of Sciences rejected the idea. Moreover, in Britain, in 1804–6, Fulton worked on mines. They were used, with scant effect, for an attack on French shipping in Boulogne in 1804, but, in trials in 1805, he became the first to sink a large ship with a mine. British interest, however, declined after Trafalgar, when Britain's naval position seemed safe; while Fulton was held back in his experiments with torpedoes from 1807 by his failure to devise an effective firing device. During the Anglo-American war of 1812–15, Fulton played a role in unsuccessful American experiments with a submarine, mines and underwater guns. He was not alone. In 1807–10, Ivan Fistum, a Russian, made advances in electrical detonation and the use of floating mines for harbour defences, while in 1809 Napoleon authorised a French company to build a submarine.[60]

Some of the application of knowledge was more mundane, as in 1775 when the British experimented with how best to ship sauerkraut to their troops in North America,[61] but this was part of a more general process of search for improvement. Yet, a focus on technology is of limited value, as, for example, is also shown by consideration of advances with balloons and rockets in conflict on land as well as sea in the wars of 1792–1815. Instead, it is more helpful to consider multidefinitional approaches to capability, effectiveness and development. Within that context, a key element was the ability to operate at reasonable effectiveness within existing constraints, and this was, and is, easier in contexts in which there are not radical shifts in technological potential, in other words periods such as 1660–1815, as opposed, for example, to the First World War. As far as the former period is considered, the key element in operating at such effectiveness tended to be institutional, a product of administrative skill, financial support and governmental stability.

All three criteria found Britain at an advantage, both before and after the French Revolution.[62] It would indeed be possible to argue that the British navy of the period 1793–1815 was so much ahead of all other navies that it could be described as a modern navy, while other navies still were early modern, although such vocabulary also introduces a problematic teleology that needs to be critically considered in terms of contemporary strategic cultures as well as the related concept of fitness for purpose in terms of specific force structures and investment patterns. To return to the concept of the British navy as modern, it was not warship technology that was different (British warships for most of the period were on the average older than those of the French), but the cumulative nature of British advances in gun foundry, gunnery methods (flintlocks), food supply and methods for storing food on ships, naval medicine and surgery, and the quality of sails and ropes. They were all considerable in 1793–1815, while they were largely absent in other navies which, to some extent, actually went backwards and were deskilled. They were certainly very different in scale to that of Britain. By 1812, the victualling system of the British navy was feeding 140,000 men daily, aside from the numerous prisoners of war who helped ensure the weakness of the French navy. This was an aspect of a formidable organisational capability that serves as a reminder of the limitations of thinking of change primarily in terms of weapons technology, with the emphasis accordingly for navies on the period 1500–1670 (including the first two Anglo-Dutch wars) and then the Age of Steam.

Furthermore, prior to the Age of Steam, the British navy was able to achieve the contemporary goals of naval power and to fulfil the assumptions of British strategic culture. As yet, the ability to act against a continental state, that was to stem from air power and, subsequently, missiles, was not conceivable, no more than the shifts that were to arise from steam power, shell guns and iron ships, or, subsequently, radio. These paradigm shifts in capability were to be revolutionary as far as ship-killing, ship control and power projection were concerned, and indeed in 1806 Fulton had argued that 'It does not require much depth of thought to trace that science by discovering gunpowder changed the whole art of war by land and sea; and by future combination may sweep military marines from the ocean.'[63] It is far from clear, however, that these later shifts in capability were as 'total' an experience of naval conflict as that during the French Revolutionary and Napoleonic wars. The length of large-scale, deep-sea naval conflict then has never been matched since, the nearest equivalent being the American–Japanese war in the Pacific in 1941–5. It is therefore paradoxical to adopt a definition of total war which excludes the naval warfare of 1793–1815. Furthermore, the standard linear concepts of military development lead to the assumption that the warfare of 1793–1815 was but a stage to a superior subsequent situation, which is not really a helpful account of naval conflict in the following decades, most of which was very limited in duration. In turn, the naval warfare of 1894–1905, the Sino-Japanese, Spanish–American, and Russo-Japanese wars, was to be different in character to that of 1816–93, but in none of those conflicts was it necessary to sustain a long-term naval struggle, unlike in the Anglo-French wars of 1689–1815.

Linearity therefore is unhelpful. Due to circumstances stemming from the period 1775–1815, particularly the sustaining then, in the face of major difficulties, of Britain's maritime strength and its ability to thwart invasion, there was no parity subsequently in the nineteenth century but rather a British naval hegemony resting on industrial strength and strategic culture which ensured that shifts in technological potential occurred within the existing hierarchy of naval power, rather than overthrowing it. This remained true even of the major challenge from Germany, a leading industrial power willing to invest in naval strength, and the assault it mounted with both surface warships and submarines during the First World War (1914–18). The subsequent shift from British to American naval hegemony, accomplished by 1945, was achieved without conflict between Britain and the USA; and even Germany and Japan in cooperation were not able to thwart this transformation. In terms of the standard linear approach, naval (and indeed imperial) rivalry between Britain and France from 1689 to 1815 can therefore be seen as part of a sequence of challenges, and not as a limited form of warfare that subsequently became total or more total.

11

SOCIAL AND POLITICAL CONTEXTS

The crippled, limping, gun-laden protagonist, dependent on begging, in *Harlequin Returning from the Wars,* a mid-eighteenth-century work by the Florentine painter Giovanni Ferretti, was as realistic an image of war as the triumphal celebrations, mingling thanks to God and man, that greeted victory. If, for most people, the state was more peripheral than the harvest and the unending struggle to safeguard crops and flocks or the incessant conflict with accidents and disease, nonetheless the most significant impact of political society was war, the damage it could create and the need to support it, through finance, recruitment and supplies. All societies were militarised, in the sense that armies were a significant government preoccupation, their financing, directly or indirectly, a major problem for both state and subject. As a consequence, war had an impact on all civilians.

The word 'contexts' in the title of this chapter is a distancing one, suggesting that social and political aspects of the subject were in some way distant from conflict and secondary to it. Analysis of these contexts in terms of issues such as recruitment contributes to this, because, while clearly important, they share this distance. This, however, is mistaken. The political dimension was absolutely crucial to the tasking of the military, while the social aspect was brought brutally home by the treatment of civilians. The period is usually discussed in terms of an extension of state control over armed force, with the control made permanent, put on a more stable financial footing and subject to tighter administrative supervision. All enabled an increased scale of war, yet (supposedly) also reduced its impact on civilians. This process is usually linked to state-building, either in terms of absolutism or of the development of the fiscal-military state discerned in Britain.

This analysis, however, has flaws. Despite claims about limited warfare and the greater bureaucratisation of armies and military life, the conduct of troops frequently pressed hard on civilians. Especially if unpaid, troops were apt to seize property from civilians and frequently to treat them, especially men, with great brutality. The harsh treatment of civilians across Europe in part reflected that which soldiers could mete out to each other. High death rates in battle habituated soldiers to killing, and at very close quarters. Furthermore, the slaughter of prisoners was frequent, especially in conflict with the Turks. This practice was readily extended to civilians, particularly in the context of religious antagonism. Military practice was frequently devastating for civilian society. Thus, the destruction of crops in order to intimidate and weaken opposing rulers, forces and towns had very harmful short-term effects on rural society. Civilians were especially vulnerable when towns were stormed. It was then conventional for troops to sack them, as much as an expression of anger and power as in search for loot. Nevertheless, towns tended not to be stormed in this period: negotiated surrenders were more common.

More generally, rulers and generals who could not afford to pay their forces transferred the cost of supporting them to the regions in which they operated. This was an important aspect of war's characteristic as, at once, destructive and also an aspect of a different tempo of state activity. Looting and enforced 'contributions' removed coin and precious metals from areas, gravely hindering their economy. Armies also caused much unintentional hardship, particularly as spreaders of disease.

The tasks of the military

Alongside war, there was also the use of troops in order to maintain control. Much violence in society was waged, handled and mediated without the intervention of the military. Nevertheless, the general absence of national police forces (and the fact that most police forces were small and under local control) ensured that when it was considered appropriate for central authorities to act, then the recourse was to the military, the largest force of men at the disposal of rulers. The role of armies in social and political control is a subject that has received insufficient attention, but troops were frequently used for policing purposes and also in support of taxation.

Furthermore, the limited nature of royal or princely control in many, especially frontier, regions was such that the army was the appropriate solution for maintaining order anyway, even without attempting the harder problem of enforcing laws on a systematic basis. The violence involved in some policing, and the size of possible problems, encouraged the use of troops, for example in sweeps against brigands or against smugglers, as by the French in Dauphiné in the early 1730s and on Belle Île in 1764. This was an aspect of the suppression of extra-territorial forms of activity particularly seen in campaigns against pirates.[1] In some states, there were more large-scale problems of control in peripheral provinces or regions, leading to the deployment of troops, for example in Catalonia within Spain. In 1664, the opposition of the city of Erfurt to the overlordship of the Archbishop-Elector of Mainz was ended by forces of the French-sponsored League of the Rhine. This was also a problem for non-Western states, for example for the Turks in Egypt, particularly in the 1780s.

Although there were fewer popular risings in the West in the period 1660–1790 than in the previous 130 (let alone 150) years, there were still many, and some, such as those in Brittany in 1675, Bohemia in 1680 and 1775, the Volga region in 1773–4, and Transylvania in 1784–5, were large-scale. A large force was deployed in Bohemia in 1775. At the same time, troops were also employed to suppress smaller-scale risings such as that in 1786 on the estate of Prince Liechtenstein on the Moravia–Lower Austria border.

Troops were also used in response to urban disorder, as in Glasgow in 1725, Madrid in 1766 and London in 1780. In the last case, the Gordon Riots, George III pressed for firm action, 'for I am convinced till the magistrates have ordered some military execution on the rioters this town will not be restored to order.' Approving the dispatch of troops to deal with the anti-Priestley riots in Birmingham in 1791, George wrote 'There is not a more indispensable duty in the executive power than to support the civil magistrates in suppressing riots, where it cannot otherwise be effected.'[2] The absence of equipment other than firearms and the lack of any training in non-violent control, which was not anyway part of the military culture of the period, could easily lead to fatalities when troops acted.

Urban disorder also led to a role for fortified positions away from external frontiers. Louis XIV built a citadel at Marseille, while, after the rebellion of Messina in Sicily was suppressed in 1678, Charles II of Spain imposed a substantial garrison in a new citadel. The Bergenhus fortress in Bergen in Norway, which was completed in the late seventeenth century, was kept operable until the start of the nineteenth century in order to control the area, a role encouraged by a local rising in 1765. In 1715, Joseph Clément, Prince-Bishop of Liège, claimed 'it is absolutely necessary that the citadel of Liège remains a fortress capable of holding the common people of Liège in awe. Without this restraint there will be no security for honest men, only murder and brigandage.' He later wrote of the need 'to prevent the evil schemes of the majority of the Liègeois who would like to become a republic and to join the United Provinces as another province . . . only troops and fortresses can hold them to their obedience.'[3]

These military roles were replicated in the colonies, as in Boston in 1768, when British troops fired on rioters. The same year, Louisiana, which was transferred from France to Spain after the Seven Years' War, also found the restrictions of an imperial trading system unacceptable. In October 1768, New Orleans rebelled against Spanish rule after trade outside the Spanish imperial system or in non-Spanish ships was banned. Having expelled the governor and his men, the Louisiana Superior Council appealed to Louis XV to restore French rule. Faced with Spanish determination to restore authority, the French, however, did not act, and in August 1769 a Spanish force occupied New Orleans and executed the revolt's leaders. Their British counterparts in Massachusetts were less successful in April 1775 when they tried to overawe opposition by seizing a cache of arms in Concord.

The previous year, General Thomas Gage, the British Commander-in-Chief in North America, had also been appointed Governor of Massachusetts.

Military power was regarded as necessary to authority as well as power. Thus, the deficiencies highlighted in the Seven Years' War led Charles III of Spain to send regular regiments for garrison duty in leading colonial centres and to set in train the raising of large forces of militia. Combined, these forces provided government with security to introduce changes, although there was resistance.[4] The captaincy-general of Caracas established in 1787 had a regiment of troops at the new capital.[5]

Troops were also used in labour disputes, French soldiers intervening against the Sedan clothworkers in 1750, while in the 1800s, in newly independent Haiti, troops were used to impose forced labour on plantation workers. French troops were also used to terrorise Huguenots (Protestants) into conversion to Catholicism in the early 1680s, as instruments of persecution until the 1780s, to enforce the cordon created around Marseille during the 1720 outbreak of plague and to enforce the new Revolutionary order in the 1790s. The use of the military to support quarantine regulations was widespread. In response to epidemics, troops were used to close frontiers in eastern Europe, as in 1753 and 1770. Venice sent warships into the Adriatic to prevent the arrival of ships from infected areas in 1743, while naval patrols were established off the Neapolitan coast in 1778.

In Russia, the police functions of the military increased. The Russian army administered and collected taxes, especially the poll tax introduced by Peter the Great in 1718, carried out censuses and acted as a police force: under Peter, Russia was divided into regimental districts in which units were quartered. Soldiers had a role in the penal system as well. More generally, the role of armies as the support of government serves as a reminder that professionalism in command, and as officers and soldiers, has to be understood within its political context, which was vital to the tasking of the military.

Troops could also be employed for purposes that bore little relation to policing activities. Officers in the Prussian cantonal conscription system used troops for work on their estates. Similarly, rulers utilised soldiers as a labour force, both for military purposes, such as the building of fortifications, and for other tasks, such as canal construction for Peter the Great, draining marshes in Prussian Pomerania, and road-building in Scotland, the last designed to serve military and economic goals. The use of soldiers for big construction projects in part reflected the fact that they were the largest group of disciplined men accustomed to working together. Armies could also be utilised for social action. The Bavarian Minister of War, Benjamin, Count Rumford, an American loyalist who experimented with the properties of gunpowder, was keen to use the army in the 1780s to introduce useful improvements. He established military gardens, with the intention of publicising new agricultural methods and crops, particularly the potato, and regimental schools that would also educate the children of the local peasantry. More generally, the strengthening of the military was seen in terms of a synergy with that of state and society, both in metropoles and in colonies.

The non-military use of troops could lead to conflict with civilians, as when garrison labour was employed in weaving or shoe repair. In some cases, as with harvesting, it could also lead to a conflict between military and civilian purposes. Furthermore, camp followers were regarded as a serious problem by civilian authorities. This underlines the multiple nature of military tasking, which, indeed, serves as a critique of the notion of a readily definable best military practice, and a resulting clear hierarchy in capability. The role of combating what was defined as disorder in society also highlights the need for discipline within the military.

The military and society

The relationship between military and society was far more complex and varied than one of simple animosity. In some respects, violence was the product not of animosity, but of a breakdown in a multifaceted set of links between troops and civilians that encompassed recruitment, billeting, logistics and sexual relations. For example, the presence of a garrison led to an increase in the rate of illegitimate births in the area. This was disruptive and also put pressure on social welfare, as extended families, poor relief, charity and orphanages had to cope with the bastards, and a large number of the women convicted of infanticide had had relationships with soldiers;[6] but the relationships should not simply be seen as adversarial. Instead, they were aspects of the situation in which the military acted and interacted as part of a complex social network, and of the extent to which the military was not separate from society, but part of it. In particular, the housing of troops was less segregated than is the case today. They were especially important in garrison towns, which included all capital cities. In Paris in 1789, a key role in the French Revolution was played by troops permanently based there, who lived and worked among the populace, pursuing civilian trades when not on duty, and were thus subject to all the economic and ideological pressures affecting the people.

The military and society each shaped the other, and the military was more shaped by civil society than is often appreciated – indeed, more than most contemporaries wanted, because this shaping corroded discipline and sapped the rigid structures of military hierarchy. As in civil society, vertical alignments existed in the military, alongside horizontal status or class-based ones, but, irrespective of these alignments, social pressures and assumptions, and the play of personality, acted to weaken the structures of hierarchy and to compromise formal discipline, although often adding the strength of informal consensual practices. There was also a feedback mechanism, with several aspects of the war–society relationship proving of particular importance for military capability and effectiveness. This was especially seen with recruitment practices and ethos, key elements in the interaction of the military and civilians. The overlap was most strongly seen with militia systems, which are underrated due to the focus on regulars, but which were nevertheless important. Militia service could bring peasants valuable status. In Savoy-Piedmont, militiamen gained legal and other privileges as a result of service, while wealthy peasants who could provide their own horses were allowed to fight duels and to hunt like nobles. Urban militias also provided a way to gain and affirm status. Furthermore, they had a military and political value, providing at least a layer of defence to all cities and sometimes the sole defence. This was particularly important in resisting the requisitions of opposing forces. It ensured that in order to obtain contributions, these forces would have to be able to overawe the cities.

It would be mistaken to adopt too schematic an approach to recruitment, not least because armed forces varied greatly, not only between peace and war, but also within individual years of peace or war. A complex interaction of opportunity and motivation, involving factors such as the reliability of pay, the season and the proximity of the harvest, affected the attraction of military service; and the situation also varied greatly by state. This included variations in tasking, not least the need to counter domestic opposition. Although regulars tended to play the key role, particularly in suppressing large-scale opposition, and their ability to do so rose with the expansion in army size in the late seventeenth century, militias also played a role, not least earlier. This also serves as a reminder of the extent to which such conflict, particularly peasant and urban risings, led to the rapid development of ad-hoc forces.[7]

These risings involved much hardship for civilians, whether or not they were involved in the risings. As a consequence, rather than thinking simply, or largely, in terms of the impact of warfare on 'civilians', it is necessary to consider that of civilians on warfare, and, indeed, the extent to which it is pertinent to treat civilians as a separate category to the military. Such a distinction might be helpful in the discussion of modern Western societies, not least as they attempt to devise and sustain norms of military behaviour, but is far less appropriate for earlier periods. As a consequence, it is necessary to be cautious about employing modern analytical categories.

In pointed contrast to the modern West, there was a willingness to accept both war and killing. In this, the military reflected the nature of society, and its actions exemplified contemporary attitudes towards human life. Killing was generally accepted as necessary, both for civil society – against crime, heresy and disorder – and in international relations. War itself seemed necessary. In modern functional terms, it was the inevitable product of an international system that lacked a hegemonic power. To contemporaries, it was natural, as the best means by which to defend interests and achieve goals. The idea that such objectives might be better achieved through diplomacy enjoyed limited purchase in a society that took conflict for granted. When, in 1693, Christian V of Denmark wished to assert his claim to the Duchy of Lauenburg, he laid siege to the fortress of Ratzeburg, an obvious step as the duchy had been occupied by Celle troops on the death of the Duke in 1689 in order to assert the claims of its duke. There was a long-standing Christian critique of unjust wars, and a related call to fight the non-believer, but the God-given nature of violence appeared demonstrated by the Bible, with frequent reference in churches to the warrior Old Testament rulers of Israel in readings from the oft-cited *Book of Kings*.

By modern Western standards, a large percentage of males served in the military, in part because the high rate of casualties ensured the need for a high rate of replacements. This increased the percentage of civilians affected by having husbands, fathers and sons in the military, and the attendant disruption and sense of loss. War therefore affected at least indirectly a large percentage of the female population.

At the same time, there was a parallel between the nature of civil society and that of the military, this constituting a relationship which was mutually supportive. For example, the social dimension of war reflected that of civil society, with an assumption that soldiers and sailors would serve at the behest, and under the control, of their social superiors, whether as feudal levies, militia, mercenaries or permanent armed forces, recruited voluntarily or by conscription. Similarly, a willingness to accept pain and privation was part of ordinary labour, and military service was another aspect of it.

Violence offered not just a parallel between military and civilian life, but also a convergence, for, rather than thinking simply of the impact of war on civil society, it is instructive to consider the implications of the bellicose nature of society, and the difficulty of isolating war, particularly civil war, from, for example, large-scale feuds or banditry. Levels of violence could also be very high for ordinary people. Furthermore, throughout society, violence was seen as the way to defend personal worth, in the form of honour, and thus seemed more acceptable. It conformed to social norms.[8]

Army size

The standard approach in military history, however, is to pretend that the state, in the shape of regular forces, monopolised force, or, at least, effective force, and that the actions of these forces constituted military history. In such a light, the relationship with society is one of the state impacting on society, with the so-called military revolution, originally dated to 1560–1660, and the absolutism held to characterise governmental aspirations and power in the late seventeenth century operating hand in hand to increase and express the power and aspirations of the state. This interpretation of the state and of its development affects both the treatment of the military and of its relationship with society, but is deeply questionable, not least because of its teleological character and its general failure to address the limitations of government power and activity and their dependence on support from the socially prominent.

Furthermore, the lessening of religious tension from the mid-seventeenth century did not prevent such tension from continuing to play a role in conflict, with attendant consequences for civilians, for example in Hungary in the 1700s. Other issues, however, came to the fore. For many states, particularly France, Austria and Russia, the century 1660–1760 saw a marked increase in the size of armies in both peace and war. The ability to strike first and hard produced obvious benefits for rulers who retained large peacetime armies or navies. Their actions were watched with concern by other powers, and only they enjoyed a real freedom of manoeuvre in international relations. This helped accentuate pressure to increase the level of military preparedness. This was particularly true, at sea, of Britain, France, Spain and the Dutch, and, later, on land, of Austria, Prussia and Russia, which were increasingly bound together in anxiety and competition, leading to alliance or confrontation.

Armies and navies retained by weaker powers, in contrast, did not compare with those of the major states, their increasing discrepancy being a significant military and political development of the post-1660 period. Peter the Great's army was far larger than the 40,000 troops in the Ukrainian army of Mazepa, and Ukraine was easily overrun in 1708; although that was not only because the Ukrainian forces were inferior in numbers, but also because Mazepa attracted little active support, and because of the quick action of Peter's commander, Prince Menshikov, in seizing Baturin, Mazepa's capital.

Further west, the growth of the Austrian and Prussian armies was not matched by that of the other German states, both cause and effect of their weakness. Hanover's attempt to challenge Prussia for dominance in northern Germany in the 1710s and 1720s was handicapped by the failure to emulate Frederick William I of Prussia's determined and successful build-up of military strength. The consequences were made clear during international crises. Confidence was hit as vulnerability was underlined. In 1740, when Hanover had a complement of only 19,422 troops, Frederick the Great of Prussia deployed nearly 30,000 troops on the Hanoverian frontier to provide cover when he led the bulk of his army to invade Silesia. The following year, an advance by 40,000 French troops into Westphalia led George II to abandon his attempts to create an anti-French coalition. The Hanoverian army continued to be in the shadow of that of Prussia, totalling 15,503 men in 1775, 23,197 in 1783 and 17,836 in 1789. In 1803, French forces were able to occupy the electorate rapidly.

Moreover, Bavarian ambitions were undermined by the small size of its army: about 8,000 in 1726, rising to nearer 12,000 by 1745 during the War of Austrian Succession. French subsidies were required to maintain greater numbers. Among other German forces, the forces of the jointly ruled electorate of Cologne and prince-bishopric of Münster were a maximum of about 7,000 in 1730 and 5,100 in 1792. In 1742, the army of the Elector of Mainz was only 4,000 strong. Many of these small forces were rarely used militarily and, unless joined with other forces, were too small to produce a field army capable of fighting a battle effectively. In 1779, John Moore pointed out with reference to the small army (fewer than 3,000 troops) of the Margrave of Baden, 'He has too just an understanding not to perceive that the greatest army he could possibly maintain, could be no defence to his dominions, situated as they are between the powerful states of France and Austria.'[9]

Rulers and the military

The existence of many small forces, however, is a reminder of the extent to which Western society was imbued with military concerns. There is a parallel with Latin America over the last half-century: militaries fulfilled functions, despite the paucity of international conflict. Armies were generally seen as an essential attribute of sovereignty. This was also linked to military aesthetics. A good-looking army was valued by rulers, and that was a reason why rulers preferred tall recruits over shorter ones. For the same reason, almost all of the rulers designed the uniforms for their troops. Red and white, the colours of the British and French, looked good even if they made soldiers more conspicuous to sharpshooters, as happened to the French at the hands of the jägers of the Duke of Brunswick's army during the Seven Years' War.

Armies were also of value for their part in enabling rulers and, to a lesser extent, aristocrats, to fulfil the roles attributed to them and which they were generally willing to discharge. Military leadership was an important role, one sanctioned by history, dynastic tradition and biblical examples. In 1788, the Austrian Chancellor, Prince Kaunitz, tried to oppose this. Urging the Emperor Joseph II to abandon the personal command of his armies against the Turks, he drew attention to examples of rulers, past and present, who left such command to their generals, 'in order to concentrate on what is properly the job of the ruler, the government and general surveillance of the state they have received from Providence.' Joseph was, however, reluctant to follow Kaunitz's advice.

Some rulers, it is true, displayed little interest in war or military affairs. Their states need to be considered in any treatment of military developments. War and military institutions tend to be studied from the 'sharp end' backwards: from fighting, via organisation and institutions, to wider impact. This tends to shift attention away from those powers that did not do much fighting. Yet, their attitudes deserve consideration, and in some cases, as with Spain and its navy under Ferdinand VI (1746–59), they could spend heavily on their forces. Ferdinand, however, concentrated on domestic matters and stayed neutral in the Seven Years' War. Portugal was at peace from the War of the Spanish Succession until invaded by France in 1807, with the exception of a brief war with Spain in 1762, and some fighting with Spain in what is now Uruguay in 1777. Denmark was at peace from 1720 until the Napoleonic wars, with the exception of a brief and unsuccessful attack on Sweden in 1788. Yet military reviews were still important for the Danish court.

Many rulers, however, saw war as their function and justification as defenders of their subjects and inheritance, a source of personal glory, dynastic aggrandisement and national fortune. The celebration of the royal hero as victor was part of a long European tradition of exalting majesty in its most impressive function, the display of power. War was not the sole sphere in which such display could occur, but it was one that best served the aggressive dynastic purpose that illuminated so many of the states of the period.

Many rulers served in person. Louis XIV accompanied the army under Turenne that invaded the Spanish Netherlands in 1667, took part in the attack on the United Provinces five years later and enjoyed commanding on other occasions. When he was not able to do so, the industrious Louis directed from afar.[10] The army that relieved Vienna in 1683 was led by John Sobieski, King of Poland, and included Elector John George III of Saxony. At the Boyne (1690), a musket ball shot the heel off William III's boot, and another shattered one of his pistols. He narrowly escaped death from a French cannon ball in 1691, fought the French at Steenkirk (1692) and Neerwinden (1693), and was in command at the successful siege of Namur (1695). James II was also at the Boyne. Victor Amadeus II of Savoy-Piedmont led his troops on the battlefield, as at Staffarda (1689), La Marsaglia (1693) and Chiari (1701) and led the crucial cavalry charge that broke through the French lines at Turin (1706). Charles XII was exposed to great danger at Poltava (1709), while Philip V was with the army that invaded Portugal in 1704, and both he and his rival for the throne of Spain, 'Charles III', commanded their forces in person for part of the war in Spain. George II was at Dettingen (1743), and in 1745, Louis XV was present when Ghent surrendered and he received the keys of the town. He was also at the sieges of Freiburg (1744) and Antwerp (1746), and the battles of Fontenoy (1745) and Lawfeldt (1747), although he was not similarly active during the Seven Years' War. Having followed Louis on his 1745 campaign, the Parisian painter Charles Parrocel exhibited ten paintings of his victories the following year. At the Battle of Coni (1744), Charles Emmanuel III of Sardinia, who displayed great courage, was opposed by Philip V's son, Don Philip. George Washington took command against Shay's Rebellion.

Emulation of ancestors, former monarchs and contemporaries, and sheer excitement, all played a role in the desire to fight. Necessity was also crucial, most obviously in civil wars, which included colonial wars of independence. As a result of these, key military figures in the struggle for independence, such as Washington and Simón Bolívar became political leaders. In Haiti, Jean-Jacques Dessalines became Emperor in 1804, while Henri Christophe, a former general, became President of Northern Haiti in 1807 and King in 1811, and Alexandre Pétion, a French officer who had become a general in the revolutionary forces, became President of Southern Haiti in 1807. In turn, the military legacy of imperial regimes were rejected, as in New York City, where the gilded equestrian statue of George III was pulled down in 1776: its metal was to be used for cartridges.[11]

The only significant exceptions were female rulers, although they could also benefit by association with victory, as did Anne of Britain with the War of the Spanish Succession, and Anna, Elizabeth and Catherine II of Russia.

Maria Theresa rallied Hungarian support politically at the outset of the War of Austrian Succession, although she did not lead them on campaign. Victories against the Turks in the war of 1768–74 gave Catherine the prestige to compensate for the questionable circumstances of her accession.

Dedication to military success was not restricted to battle. In Copenhagen in 1716, Peter the Great saw the ovens constructed to cook bread for the troops before going to the palace, while in 1722 he led his troops along the distant shores of the Caspian Sea. Peacetime supervision of the army provided rulers with excitement, a sense of mission, and an opportunity to show themselves in a favourable light as leaders. Louis XIV greatly enjoyed reviews of troops, which served as controlled displays of his power, and less powerful monarchs had similar tastes. Ferdinand IV of Naples spent much time on manoeuvres, doing so in February 1777 for the benefit of the visiting Frederick II of Hesse-Cassel, who himself enjoyed drilling troops.[12] Monarchs who did not take a personal role in campaigns could be active as heads of their armies. George III gave his views on planning for the American War of Independence,[13] although that conflict revealed the serious problem of command at a distance. This was an aspect of the extent to which warfare was not conducive to the practice of personal command, something that even Napoleon could not overcome, as the baleful consequences of his attempt to manage the Peninsular War from a distance clearly indicated. George III observed in 1799,

> It is impossible without being on the spot and seeing the face of the country that any well grounded directions can be given, farther than pointing out what is the general object wished, but how to be effected or when must be left to those who are on the spot.[14]

Looking ahead to the role of warleader, rather than general, that was to become more common in the West, George III instead associated himself with the war effort. In 1778, as the strategic situation became more dangerous following France's entry into the American War of Independence, he visited the naval dockyards at Chatham and Portsmouth, as well as the military camps at Warley Common and Cox Heath.

A definite trend towards the wearing of uniform at court can be discerned, not least by the most improbable figures, including, in Britain, the Prince Regent, later George IV. Officers of the Austrian and French armies were frequently reluctant to wear uniform in the first half of the eighteenth century, but, after 1763, Joseph II and his successors always wore uniform at court and on formal occasions. Joseph, who copied Frederick the Great of Prussia's fashion for rulers wearing military uniform, changed into his field marshal's uniform when dying. This trend was resisted only at the courts of France and Spain.[15]

If rulers combined political and military leadership, the success of a few set standards for others. Frederick the Great of Prussia, who was very successful in linking political and military roles and capabilities,[16] was particularly influential.

He led his army throughout his reign, not only winning spectacular victories (although in 1741 he fled from Mollwitz, his first battle), but also drilling the army and conducting major manoeuvres in peacetime. His approach to war was certainly not a casual one, of war as a courtly activity, a royal sport, a variation on hunting, as it had appeared in some of the literature of the sixteenth century. The chivalric notion of personal honour and exemplary reputation that had led royal commanders to place themselves in the van had been replaced by a more prudent generalship, although the brave Frederick was exposed to fire on several occasions and was bruised when a canister ball hit him in the chest at Torgau (1760). The death of a royal general, like that of Charles XII in 1718, was exceptional, but the cult of bravery led to an emphasis on royal fortitude under fire, such as George II's at Dettingen (1743). Some princely commanders and officers were killed and wounded. Charles, Duke of Brunswick was killed in 1806, as were two of the five sons of Landgrave Karl of Hesse-Cassel who served in the War of the Spanish Succession, while the future Duke Charles V of Lorraine was badly wounded at Seneffe (1674). Of the five brothers of George I of Hanover and Britain, two died in 1690 fighting the Turks, and one in 1703 fighting the French. Ernest, Duke of Cumberland, a son of George III, who served against the French with Hanoverian forces in 1793–4, 1806 and 1813–14, acquired wounds and a justified reputation for bravery, much to his father's pleasure, while Napoleon was wounded by a spent ball at the successful storming of Regensburg in 1809.

Non-royal commanders died in large numbers, including Turenne at Sassbach (1675), Montcalm and Wolfe at Quebec (1759), thirty-three Prussian generals between 1756 and 1759, and 230 of the 2,248 general officers in the French army in 1792–1815.[17] Many were also seriously wounded.[18] A demonstration of bravery was important. Sword in hand, Belle-Isle led his troops at Sahay (1742), while Saxe exposed himself to considerable danger by taking part in charges at Lawfeldt (1747).[19]

In addition to bravery, or at least fortitude, technical skill was increasingly required, a development that encouraged the publication of military literature. Frederick the Great, who carefully followed military developments and was very ready to learn from experience, saw war as a duty best discharged through training and dedication, an attitude that he sought to disseminate in Prussia. His poem 'Art of War', written in 1749, revealed Frederick's belief in the need for detailed planning and cautious execution. It was reinforced in detailed confidential works of instruction written for his officers, whom he encouraged to be aggressive and to take risks, not least by consigning unsuccessful commanders to military trials.

The nobility and the military

Military uniform was the clothing of order and obedient hierarchy. It encapsulated the participation of nobles in the service state, the systematisation of the personal links binding nobilities and monarchies. As the links between soldiers and rulers were slight, with the loyalty of the former focused on their comrades, it was the attitude of the officers that were crucial if armies were to cohere and fulfil goals. Much of the European nobility saw its role as that of military service, more particularly leadership, which validated their social and political position and provided opportunities for gain, including enhanced rank; and much recruiting was for long a matter of aristocratic officers in their areas of influence. Of the 499 nobles present at the Swedish Riksdag (Parliament) of 1693, 248 held military appointments, and military service was highly esteemed in the Swedish and Russian Tables of Ranks. As many as 35 per cent of the old Danish nobility and 17 per cent of the new were in military service in 1700, compared with only 6 and 8 per cent respectively in the civil service. In Sweden, the definition of noble privileges in 1723 reserved all senior official and military posts for the nobility, and in 1771–2, 72.8 per cent of the noble estate were or had been army officers.[20] Frederick the Great reduced the percentage of commoner officers in the Prussian army.

Although aristocratic military service was proportionately less important in western Europe, especially in England, Spain and Italy (other than Piedmont),[21] it was still of consequence. Senior military positions in the British and Spanish armies were dominated by the aristocracy, with Scottish and Anglo-Irish aristocrats being particularly prominent in the British army, and during crises there was a turning to aristocratic resources and influence. At the close of 1777, the British government resolved to raise twelve more regiments for the army at the expense of city councils and leading aristocrats who, with George III's permission, were to appoint most of the officers. In 1779, an invasion scare led the elite to take on an important role in organising volunteer support, although governmental oversight affected this process. For example, happy to approve the raising the regiment by the Duke of Ancaster, George III insisted that he only hold the rank of major as 'a steady man' was necessary for the command.[22] In Britain, as more generally, there was a tension between birth and competence. After the capture of a French frigate in 1795, George applauded the promotion of the Captain and the First Lieutenant, adding:

> as the Second Lieutenant, Mr. Maitland, conducted himself very well, I trust he will soon meet with the same favour, being a man of good family will I hope also be of advantage in the consideration, as it is certainly wise as much as possible to give encouragement if they personally deserve it to gentlemen.[23]

Frederick Maitland, the grandson of an earl, was indeed a brave officer and was to have a distinguished naval career.

Contemporaries also stressed the pressure for war from the nobility at the time of France's entry into the wars of the Polish (1733) and Austrian (1741) successions. However, the decline in the size of the French army after Louis

XIV's reign, and the general peacefulness of western Europe during 1763–89, played a role in reducing the opportunities for aristocratic service in France, with serious political consequences in terms of discontent.[24]

Under Napoleon, the identification of ruler, political elite and military activities reached a new height. It was thanks to the army and his military fame that Napoleon rose to the top. In literature and art, he projected a heroic military image as a warrior. Paintings such as Antoine Gros's portrait of Napoleon at his victory at the Bridge of Arcole in 1796 provided crucial images for the Napoleonic legend. The position of Marshal of France, abolished in 1793, was re-established in 1804, while, that year, Napoleon instituted the Legion of Honour, an award for civilian or military service that was widely given for the latter. Favoured generals were ennobled and enriched. Napoleon's Minister of War and Chief of Staff, Marshal Berthier, who became Prince of Neuchâtel in 1806 and married the niece of Maximilian I of Bavaria in 1808, had a *Galerie des Batailles,* paintings of battles in which Napoleon had fought, installed at Grosbois, formerly the seat of one of Louis XVI's brothers.

The identification of states with military activities could foster a sense of patriotism, but local privileges were vigorously protected against military infringement, particularly if they prohibited or limited recruitment, billeting and providing other resources for the armed forces. Pacifist sentiment, however, was not widely expressed. The intellectuals and religious figures who condemned aggressive war, such as the French *philosophes*, enjoyed little support. Criticisms of attempts to expand the army, such as those made in Britain or by the Dutch Patriots in the 1780s, instead, were designed to support pressure for a 'popular' army: in the British, Dutch, Württemberg and American cases, a militia. Republics and limited monarchies did tend to have small armies. This was true of Britain and Poland, and of Sweden during the Age of Liberty (1719–72), but certainly not of republican France nor of the British navy. The enthusiasm with which Sweden attacked Russia in 1741 indicates that powerful and well-armed rulers were not the sole instigators of war, although the Dutch avoided war between 1713 and 1744 and between 1748 and 1780. In the republican USA, military service provided a springboard to political success, at state or federal level.

Recruitment

A larger military led to a greater imposition on society. As this was during a period in which the European population and economy did not grow appreciably, the real burden rose greatly, while thanks also, in part, to the exacerbating consequences of economic disruption and disease, the consequences could be grim. In 1675–9, war with Sweden hit farm income hard in Denmark and Norway, and this had a detrimental consequence for public health. The burden of war led to pressures on both home countries and on foreign areas in which armies operated, in short taxation and enforced contributions. The constant revenue streams that permanent armies and navies required can be seen as crucial to a shift in governance. The impact of war and militarism on society was strengthened by this military expansion. This was particularly true of eastern Europe, where conscription, with its concomitant regulation and data-gathering, was crucial in changing the relationship between state and people. In western Europe, in contrast, there was far less conscription, serfdom and labour control.

Conscription threw much light on social politics and the wider issues of relations between civilians and the military. The willingness of governments to arm their own people is more striking than their occasional hesitation and shows that the social politics of the military was not simply the crown–elite cooperation that attracts attention. Training in the use of arms was given to what must have appeared the most unreliable members of the community, some of them criminal in their background. Crown serfs in Russia were allowed to hand over vagrants, landless labourers and other unwanted members of their villages to the army as a first stage in meeting their conscription quota. The willingness of governments to arm the poor and the marginal members of the community, rather than following the more expensive, but socially less challenging, process of hiring mercenaries, is an indication of their confidence at the time in the essential stability of the social order, as mirrored in the armed forces, and in the ability of discipline to direct action and attitudes. This was vindicated by the infrequency of army and naval mutinies,[25] although there was far more desertion and evasion of instructions, both of which are under-researched.

Recruitment represented at once the strongest pressure of the military on civil society and a bridge between the two. There was no uniformity in recruitment practices, and many reasons why soldiers would wish to serve. Recruitment could be either voluntary or compulsory, the latter either conducted in an organised fashion or quite arbitrarily. The importance of voluntary service varied considerably, but its existence reflected the appeal of the military life for many. For petty nobles, it provided a career, while for many soldiers it represented an escape from the burdens of civilian life, or the constraints of local society on young adult males. The sons of soldiers and sailors were an important source of voluntary enlistment, providing, for example, most of the recruits for the French engineers in the first half of the eighteenth century, although there was no formalised hereditary pattern of service. In contrast, in Burma and Turkey the permanent professional military under the central government was hereditary, in Turkey the janissaries. In war, these forces were supplemented by conscript levies, the resulting armies being large scale. About 200,000 men were conscripted for the unsuccessful 1785 and 1786 Burmese expeditions against Siam (Thailand).

Military service was particularly valuable for exiles, with, in Europe, large numbers of Irish Catholics going into the French army, and Scottish Jacobites being recruited for the Russian navy. However, most of the hiring of foreigners was not of exiles, but rather a response to opportunities for employment. Many individual soldiers and rulers hiring out units responded to traditional links: Scots in the service of the Dutch, Germans in that of Venice, France and the Dutch, and Swiss in that of France. In 1674, Ernst August of Osnabrück was following a well-trodden path when he agreed to provide 6,000 troops for the remainder of the Dutch War in return for Dutch subsidies. Prussia and Spain recruited many foreigners as individuals, while Britain and France hired entire units. In the early 1750s, 40,000 of the 170,000 men in the latter army were non-French, mostly from Switzerland, together with other units from Germany (the source also of British-hired units) and Ireland.[26]

Across much of Europe, conscription was conducted in an arbitrary fashion, the very antithesis of bureaucratic theory, and the limited amount of information at the disposal of the state, as well as the weakness of its policing power, ensured that conscription was less effective than in the twentieth century. In Britain, the leading naval power from the 1690s, the navy continued to be dependent on impressments by the press gang, a system that was not only arbitrary, but also only partially successful. The permanent navy consisted of ships and officers, with relatively few sailors. The formation of a reserve of seamen was proposed without result: the Register Act of 1696, which provided for a voluntary register of seamen, proved unworkable and was repealed in 1710. Subsequent proposals for legislative action met resistance. Although the enlistment of volunteers was important, the wartime navy was dependent on impressments. Although, by law, this method applied only to professional seamen, it was both abused and arbitrary. Furthermore, on many occasions, naval preparations and operations were handicapped by a lack of sailors. Possibly, however, there was no better option in the absence of any training structure for the navy, and given the difficulty of making recruitment attractive when length of service was until the end of the war. The government never seriously considered paying soldiers more; unsurprisingly so, given the size of the navy, and in light of concern over naval expenditure. The Bourbon alternative – the French and Spanish registrations of potential sailors – was not obviously superior: it led, instead, to evasion and a shortage of sailors. Moreover, political support for impressments ensured that the British navy had the manpower to sustain a large fleet.

Similar methods were frequently used in European armies, not least in the recruitment of criminals, as by the Danes in 1762. Prussian recruiters would appear at Sunday church services and seize the men afterwards. It was also common for powers to recruit troops from weaker neighbours, generally with permission, but often, particularly by Prussia, forcibly. In 1743, the Duke of Zweibrücken complained of similar conduct by France. Frederick the Great forced captured enemy troops to serve. Austria was the most significant recruiter in Germany, although it was able to rely on its position within the empire – with one exception, its ruler was the Holy Roman Emperor – to help win acceptance for its recruiting.[27]

A major consequence of recruitment practices was serious desertion, another bridging between military and civil society, although such desertion was a problem for all armies and navies, however they were raised. Of the Saxon infantry in 1717–28, 42 per cent, it has been claimed, deserted, although the rate was only about 3 per cent a year, and it has been argued that the desertion rates have been exaggerated.[28] Recruiting difficulties, nevertheless, made desertion especially serious. There was particular opposition to foreign service. In Dalecarlia in Sweden in 1743, there was a peasant rising against conscription and overseas service.

Partly in response to the drawbacks of forcible recruitment, as well as to the need for larger armies or for sustaining a certain size of force in the face of the attrition produced by long wars, various states experimented with more systematic methods of recruitment. To work effectively, these systems entailed considerable regulation of the population, which could only be achieved by means of the cooperation of the nobility. The latter was in effect the local arm of governance across most of Europe while, in addition, the bulk of the officer corps was drawn from the same group. Governments regulated and benefited from this system, and it was less dangerous for them than the older practice of nobles raising mercenary bands and directing them as they pleased. Behind the systems of conscription and the militarist patina that they gave to society, with their passes, registers, annual inspections, musters, lists and numbers painted on houses, was the constant reality of aristocratic domination of society.

Such systems increased control over the peasantry, who were less able than mercenaries to adopt a contractual attitude towards military service. The Danish ordinance of 1701, reintroducing a militia system based on conscription, tied the conscript to a specific locality during his six years of service, giving landowners opportunities to apply pressure on troublesome peasants. In 1710, the militia was used to reinforce the Danish army. In Portugal, where all men between the ages of fifteen and seventy were listed, the younger sons of each household were liable for conscription. This was unpopular and led to much desertion, which hit combat effectiveness.

The regulations underlying conscription systems reflected social norms, although the spread of conscription also altered the social politics of military service. In Russia, for example, where a system was established in 1705, the clergy were exempted, local and occupational concessions freed specified groups, and wealthy peasants could buy themselves out. The system was not extended to the Baltic provinces, Belarus and Ukraine until the reign of Catherine II (r. 1762–96). Because of the large numbers of troops that could be raised, Russia, though keen to recruit foreign officers, did not make a practice of hiring foreign mercenary units. Landlords were also expected to serve, either in the army or in government. Peter the Great's insistence on this was, however, relaxed under his successors, and, in 1762, the formal obligation to serve was abolished.

In Prussia, although volunteers remained the key element, not least providing continuity in military units, in 1693 each province was ordered to provide a certain number of recruits, which was achieved by conscription, largely of peasants. Furthermore, a cantonal system was established between 1727 and 1735. Every regiment was assigned a permanent catchment area around its peacetime garrison town, from where it drew its draftees for lifelong service. The name of every duty-bound male was entered on a roll at birth, but, due to the numerous social and economic exemptions, including Berliners and workers deemed important, such as apprentices in many industries and textile workers, most common soldiers were from the rural poor, and conscription thus served both public authority and social norms.[29] The regiments were required to be up to strength for only the few weeks of the spring reviews and summer manoeuvres. For the remainder of the year, once over the initial training period, native troops were allowed to return to their families and trades, and, even when at the garrison town, they were permitted to pursue civilian occupations. Artillery horses were also allocated to the peasantry, their care secured by inspections. The Prussian cantonal system, which was extended to Silesia in the 1740s after its conquest, and to Ansbach and Bayreuth in 1796, led to a high military participation rate among the population and to a manpower pool deep enough to allow some selectivity. It also created a stable and predictable link between regiments and reserves of manpower in specific areas that generated significant parish and regional solidarity in units, and also encouraged a sense of feudal obligation among the officers.[30]

In so far as comparisons can be made, the proportion under arms in the period 1670–1789 was comparable with that in the nineteenth century, and often exceeded it: German peacetime establishments were proportionately higher than in the nineteenth century. Even mobilisation in 1914 did not exceed the proportion mobilised by some states in the eighteenth century; it was only through the subsequent call-up of reservists that eighteenth-century totals were exceeded.

Prussia was not alone in Germany, although the degree to which most medium-sized territories retained substantial forces declined from the 1720s. In 1730, a year of peace, but also of war preparations, Hesse-Cassel, however, had one in nineteen of the population under arms. In 1762, Frederick II of Hesse-Cassel divided his territory on the Prussian model into recruiting cantons, one for each regiment. Recruiting by force was prohibited, and exemptions given, either by payment or by occupation. The major towns were exempt, as were propertied farmers, taxpayers, apprentices, salt-workers, miners, other important workers, domestic servants and students. Eligible men were listed and checked annually.[31]

Like Prussia and Hesse-Cassel, Sweden used a cantonal system, with, in addition, from the 1680s, farms in royal hands used to support soldiers who worked there in peacetime. Regular training and periodic musters were designed to keep the troops effective, and, as a result of the successful operation of this system, Charles XII inherited a well-trained army of 90,000 men in 1697. After his death in 1718, however, the army deteriorated. There was a lack of political will for a strong army in the Age of Liberty, and more of an emphasis in the royal farms was placed on farming.

Uniformity in the Habsburg military system was lessened by the privileges of particular areas, the Habsburgs encountering fewer difficulties outside Hungary. A reserve system was introduced in 1753, and short-service voluntary enlistment in 1757. In 1771, Joseph II brought in a system of conscription to Austria and Bohemia, by which each district was expected to raise a certain number of troops, although there were important exemptions, including those who were most economically productive. Thus the army, which in 1789, was over 300,000 strong, was largely a force of poor peasants. The artillery, however, was a more select force, composed of volunteers able to read and write German, and therefore more easily trained for a higher level of skill.

Conscription was imposed in Spain in 1704, an example, like the kingdom of Sardinia in 1744, of its being introduced when there was a national emergency. Thereafter in Spain, however, the defects and unpopularity of conscription ensured that it was only used when there were insufficient volunteers. Until 1776, when evasion and unpopularity led the Spanish government to abandon it, conscription was by ballot among unmarried men aged seventeen to thirty-six, although with many exemptions, including skilled workers. This was supplemented by the levy by which convicts, beggars and vagrants could be conscripted. After 1776, there was a reliance on Spanish and foreign volunteers, but this did not provide the numbers required.

In Britain, which lacked a large peacetime army, conscription was regarded as unacceptable, although there were impressments of the unemployed during some periods of acute manpower shortage, such as the War of the Spanish Succession (1702–13). In 1756, when the Seven Years' War began, compulsory enlistment was made possible by an act of parliament, but this was deemed a failure: it proved difficult to raise sufficient men, their quality was low, and desertion was a major problem; and the system fell into disuse in 1758. Britain lacked a regulatory regime and social system akin to Prussia or Russia and, without them, it was difficult to make a success of conscription or to control desertion. In 1780, when Britain was under great pressure, Charles Jenkinson, the Secretary at War, wrote to Jeffrey, Lord Amherst, the Commander-in-Chief, that he did not see how the strength of the army could be maintained, but he added,

> I am convinced that any plan of compulsion . . . is not only contrary to the nature of the government of this country, but would create riots and disturbances which might require more men for the purpose of preserving the peace, than would be obtained by the plan itself . . . besides that men who are procured in this way almost constantly desert, or at least make very indifferent soldiers.[32]

The peacetime British army was only about 30,000 strong in the first half of the century, and 45,000 strong in the 1760s, but, in wartime, it was possible to use militia and other volunteer forces for home defence, as well as to expand the regular army. The hiring of forces was also an option, as was reliance on the armies of allies, some of which were subsidised.

There was also a reliance on forces raised within the imperial system: colonial militias in the New World, the large sepoy (Indian) army raised by the East India Company, and also the black auxiliaries used against escaped and rebellious slaves in Jamaica in the 1730s and 1760. In 1741, Etienne de Silhouette, a French agent in Britain, in a letter intercepted by the British, reacted with alarm to the news that the British were arming black men to use them against Spanish-ruled Cuba. He felt this dangerous for all Western empires, but argued that the British were too obsessed by their goals to consider the wider implications. Concern about the views of local white people in the West Indies, as well as the hostile response to the arming of Virginian black men in 1775–76, led to little British support for the arming of black men during the American War of Independence, although the French colonial militia sent to help the Americans in Georgia in 1779 included free black men from Saint-Domingue. In the 1790s, in response to the French arming black men in the West Indies, the British did the same. During the war of 1812, the British willingness to receive and arm escaped slaves aroused American anger.[33]

Native French forces were raised voluntarily in the case of regulars, with both the personal contacts of officers and recruiting parties playing key parts, although force could also play a role. The recruitment of volunteers was handled by officers, generally nobles, who recruited in their seigniory of origin. From 1688 conscription among unmarried peasants aged eighteen to forty was used to man a royal provincial militia designed to provide low-cost auxiliary troops. This ended with peace in 1697, to be revived in the War of the Spanish Succession. From 1703, married men were also conscripted. From 1693, militia units were sent to fight in war zones, and, from 1701, the militia merged into regular regiments. In 1706–12, this compulsory transfer was achieved by the drawing of lots, which helped to make militia service very unpopular. Abolished in 1715, but resumed in 1726, to be a permanent part of the *ancien régime* military system,[34] militia service was affected by exemptions and these were maintained vigorously. The 1743 proposal to extend the militia to Paris led to widespread concern and the production of critical handbills, but, nevertheless, it was extended there two years later. Exemptions for military service were permitted on occupational grounds to the benefit of those employed in trade, manufacturing and public function. Very few peasants were exempt.

Peasant hostility to recruitment, however, which, in part, reflected a sense of soldiers as rootless individuals lost to the land,[35] helped ensure that a disproportionate number of recruits for the army were urban poor, especially from Paris. The army was young: on the eve of the Revolution, half of the infantry of the line were aged eighteen to twenty-five, and 90 per cent of soldiers in all arms were aged thirty-five or younger. The majority of those serving in 1789 had been recruited as recently as 1783–5; only a fifth had ten or more years' service.

Although desertion was a major problem, the French system worked well in mobilising large numbers for the army in time of war. In the early 1750s, there were 130,000 French soldiers, slightly more than 0.5 per cent of the population (as well as 40,000 non-French troops). During the Seven Years' War (1756–63), men under arms, including the navy, at any one moment totalled 2.5 per cent, while the need to replace losses ensured that in total nearly 1 million Frenchmen served, about 4 per cent of the pre-war population. This was achieved with relatively minor harmful economic effects, suggesting that recruits were men whose labour could be reallocated with least difficulty, thanks to the significant underemployment of the rural population.[36] For some, military service can be seen as a form of migration comparable to that of movement to the towns. Thus, the French failure to establish a conscription system during their *ancien régime* identical with that of Prussia or Russia reflected the success of their own methods, which itself was a testimony to the size of the population of western Europe. This was further underlined by the ineligibility on health grounds of much of the population. The French army drew its troops from the sixteen-to-forty age group, recruiting first among unmarried males: between a quarter and a half of this group were ineligible because they were infirm or too short, the latter often a product of poor nutrition.[37]

Revolutionary and Napoleonic France, however, required far larger numbers of recruits. As with the Americans in their War of Independence, it swiftly became clear that volunteers alone would not suffice. There were too few troops in 1793 for the challenges confronting the Revolutionary government, and this led to the *levée en masse*. Although the reality of the resulting system did not measure up to aspirations, this was akin to later systems of mass conscription and it helped raise land warfare to a new level of scale (see p. 119). The mutual impact of warfare and growing national feeling, both in France and elsewhere, was crucially linked to the expansion of conscription. Conscription and desertion became more obvious features of the social impact of war, and opposition to conscription contributed to general war-weariness.[38] This also affected areas conquered by, annexed to, and allied with France.

Territorial control was directly linked to military manpower. Thus, the Austrian Netherlands, which had not had conscription, provided nearly 100,000 men between 1798 and 1809, when annexed by France. In 1797, the French army fell to 365,000 effectives, but in 1798, under the Jourdan-Delbrel law, the obligatory nature of military service in France was extended: 400,000 new soldiers were raised to face the Second Coalition. The new system was designed to produce a more stable system than the *levée en masse*. Subsequently, the Napoleonic military machine owed much to the troops provided by allied powers. French control over them was often direct: in 1809, the Saxon army was put under the command of Marshal Bernadotte and designated the IX Corps of the Grande Armée. Conscription was introduced by Napoleon's ally Bavaria in 1804–5 and extended there in 1809 and 1812. The French supported their presence in Italy by conscription, so that in 1798 it was decreed that all males between eighteen and twenty-six be enrolled.[39] In 1809, Marshal Joachim Murat, the new French ruler of Naples, followed suit. In France and Italy, the writings of soldiers indicated that some of them were motivated by ideological factors. When the French occupied Portugal in 1807 they raised 14,000 men, and the Tyrolean rebellion of 1809 was in part a result of the introduction of conscription. When Napoleon annexed the kingdom of Holland in 1810, conscription was introduced: a very unpopular measure.[40]

Within France the burden of conscription, however, became insupportable from 1812, especially in 1813, when 900,000 conscripts were called up, and evasion and desertion became far more serious, especially in the Auvergne and the south-west. The larger numbers involved in Napoleonic warfare ensured that conscripts played a greater role than in pre-Revolutionary conflict, in which volunteers and mercenaries had played a proportionately larger part. Everywhere, the search for more soldiers and sailors were stepped up. In Austria, exemptions for recruitment were tightened up in 1804, although the Hungarian Diet rejected conscription and the militia. In Norway, the war between Denmark and Britain in 1807–14 led to the enlistment for the coast guard of men aged from eighteen to fifty-five within 10 kilometres of the coast, and they manned the over 100 fortified positions which were prepared.

The social contours of military service are generally treated in broad-brush terms, but it is probable that, just as elite behaviour was affected by individual preference, factional politics, and family traditions and strategies, so there was considerable variety within the groups sometimes undifferentiated as soldiers and peasantry. The regional variations in recruitment found in France would have been seen elsewhere, while the complex web of assumptions and rituals that produced unit cohesion would have been constrained by, as well as made possible, the exercise of authority, and also ensured considerable variety. This is an instance of the degree to which war and the military were both distinctive and yet also shared in wider social and political processes, especially shifts in the relationships between socio-political groups. This has to be grasped if the intellectual challenge of the subject is to be confronted.

Military life and conditions

The relationship between the military and wider social and political processes can be seen if the conditions of military life are considered. The need for drill, the attitude towards soldiers of most officers (a group with whom they had little social affinity), and the frequency of desertion, led to a discipline that was often harsh or hostile. The limited effectiveness of weapons ensured the need, both for troops and warships, to come close to the enemy in as large a number as possible. This was probably a frightening experience: the enemy could be seen clearly, the noise of the battlefield was tremendous, and casualty rates were often high. The complicated loading drill for the weapons of the period required the conditioning of soldiers to repeat them under stress and the effect of weapon fire depended on the standard of fire discipline. Drill and discipline were therefore military necessities, on both land and sea.

Peacetime conditions were far less dangerous, although the desire to prepare troops for future conflict was not always matched by care for them. The debilitating effect of peace was stressed frequently by commentators, particularly military ones who disliked peacetime economies and the lessened stress on training. Peacetime life could certainly be boring for soldiers; it was difficult to maintain weapons and uniforms to the standards judged necessary; and insanitary living conditions encouraged poor health and disease. Pay was another problem and often affected morale. Soldiers were a marginal group in society, affected by the general antipathy of settled communities to the army as 'outsiders', although this exclusion was tempered by the large number of troops who pursued civilian trades. The very slow spread of barracks, an expensive device, indeed ensured that troops were frequently quartered among the civilian population, a policy that made it harder to maintain discipline and train troops, and even to develop a distinctive military attitude (see p. 44). The use of barracks became more important in Austria, Britain, France and Prussia in the second half of the century, while it was decided to house all Moscow's soldiers in barracks in 1765. The Royal Barracks in Dublin had been constructed in 1705–9. Barracks, however, were often unpleasant for the soldiers, frequently being gloomy, damp and insanitary.

It is important not to exaggerate the miseries of military life. The military were a section of the community which governments needed and cared for, albeit at a basic level. Their basic sustenance was provided during peace, although that was scarcely true in war. In the 1720s, every French soldier in Roussillon received daily about 1 kilogram of bread, 500 grams each of vegetables and meat, 25 grams of fat, and 1 litre of wine or beer. From 1803, the Batavian (Dutch) Republic supplied each soldier daily with 1 pound of bread, the basic minimum that virtually all governments seem to have provided. Though pay was generally low and frequently delayed, troops were the largest group paid by governments.

Training, in addition, seems to have created a bond, generally not only between soldiers but also between troops and officers. This was especially strong in the cavalry. The good relationship between officers and men in the French cavalry was because of the paternalistic values of the officers and the high status of the ordinary cavalry trooper. Cavalry (except dragoons and hussars) were generally better disciplined. They attracted a higher quality recruit (virtually all volunteers), who received higher pay.

Discipline in general was not always as savage in practice as it was in theory, a common feature of the law enforcement of the period, which was often tempered and episodic. The Prussian army has a tough reputation, but the most vivid accounts of the system were selective treatments composed by critics. A small number of hard cases received a disproportionate number of the most severe punishments, and Frederick the Great used the death penalty very sparingly. Nevertheless, defeat at Rossbach (1757) was followed by the introduction of stricter discipline on the Prussian model into the French army. Russian discipline appears to have been more savage, possibly due to the practice of treating soldiers as possessions of their officers and the fact that service divorced them from their native communities.

Alongside discipline, there were signs of improvement in the treatment of soldiers, although it is unclear how far these were linked to humanitarian change in civil society and the advocacy of enlightened writers within and outside the military. A growing emphasis on the value of the motivation of individual soldiers was doubtless of consequence in their treatment; and medical and hospital provision improved from the late seventeenth century. Russian regulations for officers issued in the 1760s emphasised the need for positive motivation in the process of transforming peasants into soldiers. Utilitarian uniforms, which were easier to maintain, were introduced for the Austrian infantry in 1767. In the 1780s, Rumford improved the pay and conditions of Bavarian soldiers, and in 1764 France created invalidity pensions. Ex-soldiers were provided with a pension and allowed to live in a residence of their choice, the first true attempt by the monarchy to provide old soldiers with the possibility of gaining an honourable place in society. There were, however, serious limitations. French policy towards veterans, and the administration of the Hôtel des Invalides in Paris, founded by Louis XIV to care for them, displayed paternalism, favouritism and the effects of social privileges; reforms and financial expedients were inconsistently applied; and occasional royal visits and the distribution of largesse could not satisfy the needy veterans. The Invalides catered only for a minority of officers and men, and it was not until the pension law of 1790 that all veterans of the same rank received equitable treatment based on length of service.[41] However, pre-Revolutionary policy towards ex-soldiers did reflect humanitarian considerations.

These were also displayed in the French navy in the 1780s. In 1781, pay and bonuses were increased, travelling expenses for recruits introduced, and the decision taken to award half-pay to injured novice sailors who could no longer earn their living. In 1782, the government assumed the responsibility for lodging naval conscripts before their assignment and also began work on a naval hospital at Toulon. In 1784, a regulation dealing with food for the navy insisted that the flour should be of the finest quality, the wine of good stock and that the sailors were to receive pork and beef; while the size and regularity of pension payments were improved, and naval discipline was eased. In 1786, Castries projected an increase in the number of hospitals, as well as higher salaries for doctors and surgeons, and he raised the overall number of medical officers in the ports. The ordinances of 1786 reflected the fruitful fusion of regulation and science that characterised some of the most progressive legislation of the period. Castries issued specific orders for washing down the ships with cleansing agents. He established a standard uniform for the first time and took steps to ensure that ships were equipped with medicines and with foodstuffs necessary for recuperating sailors. Castries also instigated research by the Royal Society of Medicine in Paris on the preparation and preservation of foodstuffs on board ship, the nutritional requirements of seamen, ship ventilation, and the treatment of a number of illnesses.

However, there was little time to implement Castries' programme before the Revolution.[42] Indeed, many *ancien régime* reforms in the treatment of soldiers and sailors, such as the Dutch measures of the early 1730s to improve conditions for sailors, or those of 1741 to limit typhus, were of scant effect. In his last illness, Frederick the Great declared, 'In the course of my campaigns all my orders relating to the care of the sick and wounded soldiers have been badly observed.'[43]

More generally, reforms demonstrate the limitations of the idea that the position of troops, any more than that of peasants, was uniform, or that their treatment was always bleak. The fighting qualities of the armies and navies that confronted the forces of Revolutionary France did not stem simply from discipline or belief in their cause. They also reflected a professionalism born of training and responsible treatment. Discipline was enforced within the context of cooperation between officers, non-commissioned officers and men. A sense of self-value as a soldier played a role among Prussian troops, as did education and exhortation by regimental chaplains, who found service in the army a valuable source of appointment and promotion.[44] French soldiers, who, as a result of the Revolution became citizen-soldiers, also displayed very varied responses ranging from patriotic enthusiasm (the minority) to professional cohesion, although homesickness was a widespread problem, and one that encouraged desertion.[45]

Popular warfare

Ancien régime warfare is often presented as civilised and limited, certainly in contrast with the previous period of religious zeal and the subsequent one of political enthusiasm. This is questionable as a description of conflict between regular forces and also ignores the extent to which the civilian population could be directly involved, for example with the harsh exaction of supplies combined with scorched-earth policies, as by the French in the Palatinate in 1674 and 1688–9. Guerrilla action played a role in conflict in the period. The Swedes who overran Courland in 1658 met serious guerrilla warfare, with peasant units mounting a number of counter-attacks, one of which temporarily overran the western suburbs of Riga in 1659.[46] Although the scale varied, regular troops were harassed by guerrillas in Piedmont, Dauphiné and Spain in the 1690s, and Hungary and Spain in the 1700s. Indeed Jean-Jacques Pelet, first aide-de-camp to the Napoleonic Marshal Massena, was to draw comparisons between the War of the Spanish Succession and the Peninsular War. In 1690, peasant retaliation against French exactions in Piedmont led to a systematic sacking of the area.[47] In the 1700s, there was intensive Polish guerrilla warfare against the Swedish invaders in 1703–4, while in 1707 Swedish demands for food in the Masurian Woods region in Poland led to guerrilla warfare.

Guerrillas also played a considerable role in Iberian operations during the War of the Spanish Succession, although not on the scale of the Peninsular War of 1808–14 because then the French throughout Spain and Portugal were in effect fighting in hostile territory, not that all the guerrillas were patriots.[48] Marshal Berwick, who led the Franco-Spanish invasion of Portugal in 1704, was surprised by the weakness of organised resistance, but equally amazed by the vigour of the peasants in attacks on his communications and in fighting back in the villages. Their success in exacerbating his supply problems played a major role in inducing Berwick to retreat. Indeed, guerrillas were used by both sides during the operations in Spain, and the response from regular forces was generally harsh. The French killed the survivors when Xativa fell in 1707 and left no building standing except the church.[49]

Soldiers were not trained for counter-insurgency warfare, and many reacted by destroying the homes and crops of peasants in order to wreck their fragile economy. Examples included the Protestant rising in the Cévennes mountains of southern France in 1702–11,[50] Tyrolean peasant resistance to a Bavarian invasion in 1703, the subsequent Bavarian peasant rising against Austrian occupiers, the suppression of the Jacobite Highlanders in Scotland in 1746, the Genoese popular rising that drove out Austrian occupiers in 1746, and resistance in Corsica to French conquest in 1768–70. In the last case, the French responded to guerrilla-type resistance with devastation and terror, as well as by building roads in order to improve their capacity to respond. In Genoa, the rising against Austria had led to partisan warfare by local peasants, the training of priests to fight, and women working on the fortifications.

Guerrilla warfare is not always easy to define, but its existence calls into question the habit of typecasting eighteenth-century military operations in terms of the predictable combat of regular units. Such warfare also directs attention to popular political consciousness, and to the extent to which international and internal conflicts could overlap and interact.[51] If guerrilla action was often provoked by the exactions of hostile soldiers, that does not mean that they necessarily defined this consciousness. Nor was it only a case of guerrilla warfare. Popular forces also engaged in battle, as in the unsuccessful attack on loyal troops at Sedgemoor in England in 1685. In 1708, a popular revolt delivered Bruges and Ghent to the French, while, in the Swiss Civil War of 1712, Zurich had 20,000 of its citizens under arms. Two years later, Barcelona was defended with popular fervour against overwhelming Franco-Spanish attack. A worker-militia manned the walls and resisted several attacks successfully. Popular enthusiasm was sustained by religious fervour, including the use of sacred relics. Although the bombardment began on 11 May 1714, and attacks on 25 July, Barcelona did not fall until 13 September of that year.[52] Thus *ancien régime* warfare should not be divorced from a popular context in order to contrast it with earlier and later periods. Popular participation in the Revolutionary and Napoleonic period was doubtless more sustained, but it was not unprecedented.

The treatment of civilians

Punitive violence was directed at civilians who took part in military activity. This was abundantly seen in the Scottish Highlands in 1746 as loyal troops suppressed Jacobite action, although the Jacobites themselves could also be harsh. In 1716, during an earlier uprising, in a fruitless scorched-earth attempt to delay John, 2nd Duke of Argyll's march, the Jacobites had burnt down settlements such as Crieff. In 1746, the Scottish Highlanders who provided the core Jacobite support were regarded as barbarians and, indeed, compared unfavourably with the Gauls by Thomas Ashe Lee, a British army officer who sought an historical comparison to the campaign by reading Julius Caesar. William, Duke of Cumberland, the commander of the army, allowed his soldiers to plunder Jacobite property on the march north, and his actions were criticised, especially in loyal Scottish circles. John Maule MP wrote from London,

> There are many here who much disapprove of the plundering that has been but too much practised by the Duke's army, no good can result from it, and the rebels will certainly make reprisals a hundred fold, so that the whole country will be a scene of blood and rapine and many a man that has merit with the government undone. In the meantime it is not lawful for any man to complain lest he be taxed with disaffection.[53]

The Duke regarded such action as effective, writing

> The orders which were sent to the Governor and Commandant of Fort William, before it was threatened with a siege, to seize all the cattle and demolish the habitations of those in Lochaber, who were actually out in rebellion, has had a very good effect, as all the rebels of that country have deserted, to go home to their own habitations ... which will only discourage the men and add to their present distraction.[54]

Cumberland, however, did not resort to systematic violence on a massive scale until after his total victory at the Battle of Culloden when he could spare the troops for punitive expeditions and it was safe to do so. These expeditions were ordered to kill the Jacobites and to destroy their property. Cumberland's secretary, Sir Everard Fawkener, wrote that 'persons sheltering or concealing rebels should no doubt be treated as such', adding of Lochaber 'these hills will now have been thoroughly rummaged, and the inhabitants will have learned that they have placed a vain trust in them.'[55] Some of the expeditions, particularly those sent into remoter sections of the western Highlands, were especially cruel, characterised by killings, rapes and systematic devastation that did not exempt loyal Highlanders. Those who complained were not heeded by the Duke, and were often abused. The situation was probably exacerbated by the extent to which much of the army disliked having to operate in Scotland. Even if Highlanders could flee the approach of troops and warships, the destruction of their homes and farm implements and the seizure of their cattle were for many, especially the weak, equivalent to sentences of death or at least severe hardship and malnutrition. Reports of Jacobite atrocities had lessened whatever reluctance there might have been to punish the Highlanders, who were commonly presented as subhuman, although the cruelty that was inflicted varied in its intensity, and some honourable men applied their instructions in a favourable manner, thus earning criticism from Cumberland and his acolytes.[56]

These savageries were far from unique in the eighteenth century, although they were also far from normal. The Austro-Turkish war of 1683–99 was not as cruel or savage as often claimed. Nevertheless, central and eastern European wars were often brutal, with ethnic and religious differences separating civilians from troops, and the search for supplies presenting a considerable problem. In 1742, an officer with the Austrian army reported of the attacking Prussians and Saxons,

> we all see the barbarous and inhumane effects of their visit in Moravia, and never was the like seen or heard of, and it is really scandalous beyond all measure that first towns and villages should be strongly taxed under contribution on arrival, and on departure pillaged and burned to the ground; this one may call the cluster or rabble of piratical incendiaries and not the generous valiant behaviour of a Christian army.[57]

Attacks upon civilians were even more common further east, with little distinction being observed between soldiers and civilians. The Ukrainian administrator Samub Velychko wrote of the Ukraine and Galicia in the 1700s, 'I saw many towns and castles empty and deserted, and the walls, once constructed by men to resemble hills, now serving as the homes and refuge for wild beasts.'[58] The Muslim population of Montenegro was massacred by the Serbs. When Prince Cantemir of Moldavia rebelled against the Turks in 1711, he called the entire adult population to military service, and when the rebellion was suppressed many fled to Russia. Guerrilla warfare was a marked feature of the Balkans. The attempt by the Russians to instigate Greek risings during their 1768–74 war with Turkey led to Turkish atrocities that were widely reported in Europe. Ethnic differences also played a role in rebellions in China, for example the Chin-Ch'uan tribal rising in Szechwan in 1746–9 and the Yo tribal rising in Kwangsi in 1790.

Western European operations were by no means free from savage fighting or atrocities. Henry St John, a Tory MP, complained that English conduct during the unsuccessful attack on Cadiz in 1702 during the War of Spanish Succession, which, in Spain, was a civil war, had jeopardised their chances of winning support in Spain, 'Huns, Goths or vandals never proved themselves such barbarians as we have done. Neither saint, nun, church, or convent were spared.' A British commentator wrote of the unsuccessful French forces in Germany in 1743, 'We hear of sacrileges, murders, rapes, and all the acts of people exasperated by despair.'[59] Four years later, after the French stormed the great Dutch fortress of Bergen-op-Zoom, they gave the town over to massacre, rape and pillage. The looting of the French port of Cherbourg by drunken British troops in 1758 did little for their reputation. People fled from war zones, as from the Rhineland to neutral Denmark in 1761, scarcely suggestive of any confidence in the limited nature of warfare, and indeed the routine short-change of military activity could be devastating. Even more so were such practices as the bombardment of fortified towns, such as the French use of red-hot cannonballs, designed to cause fires, against Brussels in 1695. The French bombardments of Mons and Liège four years earlier were also very damaging. Transoceanic conflict could also be very harrowing. The French population of Acadia (Nova Scotia) were deported by the British as a security risk in 1755, to be followed, after its conquest, by the population of Île Royle, renamed Cape Breton Island. In India, the French base of Chandernagore was totally destroyed by the British having been captured in 1757, as in 1761 was most of Pondicherry, including the churches, the public buildings and the fortifications; furthermore, the population was ordered to leave.

The effects of war were long-lasting in combat zones. Travelling in 1764 through the republic of Genoa, where there had been heavy fighting in 1746–8, John Holroyd, later Earl of Sheffield, noted 'the Croats of the Austrian army in their retreat from Genoa through this country burnt and destroyed all the houses etc . . . The Croats had even killed women and children. The peasants were in arms.' John Moore wrote in 1779 of the pillaging of the Palatinate ninety years earlier, 'the particulars of that dismal scene have been transmitted from father to son, and are still spoke of with horror by the peasantry of this country, among whom the French nation is held in detestation to this day.'[60] Nevertheless, the situation in the Highlands was more comparable in the level of punitive action to eastern than to western Europe, unsurprisingly so in light of the strength of hatred, the fear engendered by the near-success of the '45 and the difficulty of distinguishing between soldiers and civilians.

During the American War of Independence, in contrast, the British government and army were more cautious in their treatment of the American patriots, despite the fact that they were legally rebels. Raids on rebel-held towns, such as the one by the amphibious force that destroyed Falmouth (now Portland, Maine) on 18 October 1775, created outrage, but they were not typical of the conflict. Similarly the propaganda use the Americans made of the scalping of Jane McCrea by Indian scouts working for the British in 1777 created a false impression, although such propaganda was seen as important to the stiffening of resistance. It also reflected normative values on the nature of acceptable violence. The Declaration of Independence signed in 1776 complained that George III 'has endeavoured to bring on the inhabitants of our frontiers the merciless Indian savages, whose known rule of warfare is an undistinguished destruction of all ages, sexes, and conditions.'

In fact, the generally cautious British approach reflected the politics of the war: the restoration of the colonies to royal government would be of limited value if subsequently holding them down required a substantial garrison and if the embers of rebellion remained among a discontented population. In contrast, conflict between patriots and loyalists could be far more vicious, and was so, in particular, in the South in 1780–1. Local patriots considered the loyalists as rebels against the legitimate government and so justified their inflicting the harsh treatment appropriate for defeated rebels or in accord with the 'law of retaliation.' One retaliation naturally led to another, particularly in an environment where so much of the military activity was carried out by independently operating and institutionally weak militia forces.

Nevertheless, there was a contrast between the treatment of civilians (and prisoners) in Scotland in 1746 and that in the American War of Independence, although in each case the British were responding to a rebellion. This indicates the extent of variety that subverts any single or simple account of the subject. If an overall conclusion is to be offered, it must be one that stresses the problematic nature of the concept of the civilian, and also the questionable nature of the argument that pre-French Revolution Western warfare was limited. For the many who suffered, both military and civilians, this was certainly an age of total war. Scale and commitment were also each seen prior to 1792. At the same time, the French Revolution–Napoleonic period was understandably noted for both until fresh conflicts came to overlay their memory.[61] The most dramatic episode was the burning of Moscow in 1812, followed by the destruction of most of the French army on its retreat from Russia, but there were also other signs of intense struggle. These included large-scale long-term imprisonment. The flavour of attitudes can be glimpsed from a memorandum of 1796 by General David Dundas, a leading British general and exponent of Prussian tactical methods, outlining the intended response if the French invaded Britain, as indeed seemed a prospect,

> When an enemy lands, all the difficulties of civil government and the restraint of forms cease; every thing must give way to the supplying and strengthening the army, repelling the enemy ... the strongest and

most effectual measures are necessary . . . The great object must be constantly to harass, alarm and fire on an enemy, and to impede his progress till a sufficient force assembles to attack him . . . every inch of ground, every field may to a degree be disputed, even by inferior number . . . The country must be driven, and every thing useful within his reach destroyed without mercy.[62]

12

CONCLUSIONS

The theme of this book is that of variety in circumstances and diversity in responses, with this multiplicity encompassing social-political contexts, military tasks and methods of war-making. This entails downgrading the centrality of technology and replacing the conventional Whiggish and teleological conceptual framework for military history with an emphasis on tasks and change as problems to overcome. Strategic culture is a helpful concept because to understand the military preparations and warfare of the period, it is necessary to appreciate the great diversity in international power politics and geography.

The questioning of earlier conceptual views, with their emphasis on an ideal state of *ancien régime* warfare and a clear-cut nature of Revolutionary change, now needs to be amplified by detailed work on conflicts hitherto somewhat neglected, such as the War of the Polish Succession (1733–5) or conflict in Poland in 1768–72 prior to the First Partition. If this leads to an assessment of the period that further underlines its diversity that will be a welcome invitation to approach earlier and later periods in the same light.

Furthermore, the military history of the period will change as research extends, in particular to take advantage of the opportunities opened up by the fall of the Iron Curtain. Generally seen as a completed political development, this in fact is an incomplete cultural project, as the long-standing tendency to regard eastern Europe as primitive remains all too potent.[1] The impact of the problems that western European scholars face with eastern European languages may well be lessened as research flourishes in eastern Europe and is presented to the outside world in English. Topics that could be probed include the Saxon and Polish campaigns against the Turks in the 1690s, and the War of Polish Succession in Poland itself, a subject that has been generally neglected. Moreover, from the 1740s until the 1790s, the Prussian army enjoyed the highest reputation in Europe, but this has led to a marked undervaluation of the achievement of the Russians, an aspect of a more general relative neglect over the past 400 years. The Russians demonstrated marked fighting quality, unit cohesion, discipline and persistence on the battlefield against the Prussians in the late 1750s. Furthermore, in their wars with the Turks in 1768–74, 1787–92 and 1806–12, the Russians went on to display flexibility and to win success on both sea and, more lastingly, on land.

Research on developments in eastern Europe (and indeed outside Europe) needs to abandon the conceptions (and the vocabulary that appears to make them normative) framed on a western European teleology, not least by addressing the extent of 'sub-state' violence, and thus the extent to which internal politics were not demilitarised. For example, the imperial character of states in eastern Europe was matched by an often only limited control over frontier regions, such as Transylvania and Ukraine, especially in the late seventeenth century, and it is necessary to devote appropriate attention to their military history. More generally, across the West and further afield, it is also important to consider elements that cannot be so readily seen as 'proto-states'.[2] An account of military activity that gives weight to warriors who were only imperfectly within state structures, and indeed were sometimes classed as brigands, can be matched by an (overlapping but not co-terminous) reconceptualisation of war so that it is not simply seen in terms of state-to-state conflict, or of civil wars between forces that can be treated in these terms. Instead, the multiple concepts, causes and consequences of violence repay consideration, not least so that the social dimension is understood not in terms only of war and society but, more generally, as violence and society. Civil society might have been increasingly demilitarised across much of the West, but there were still levels of violence that created problems for the authorities.

War is often seen as a forcer of innovation: technical in the form of new weaponry; tactical, organisational and doctrinal, in order to gain advantages over opponents; and governmental and social, in the shape of the demands created by the burdens of major conflicts. Such pressures were not new to this period, while patterns of causality can anyway be difficult to establish, but the degree of administrative sophistication required to sustain the large-scale Western military systems of the period was considerable. The necessary financial and logistical mechanisms, moreover, created multiple pressure points. These tend to be approached in terms of 'inputs', the provision of resources, and related political and administrative demands, and their consequences for state development, but 'outputs', in terms of the management and utilisation of resources by the military, were also a major issue. Deficiencies in military systems were readily apparent during the conflicts of 1792–1815, but the military-administrative capabilities of the states of this period, nevertheless, were greater than they had been three centuries earlier. This was not the consequence of a smooth progression, but, rather, of accelerated development, in both qualitative and quantitative terms, from the late seventeenth century, particularly in Austria, Britain, France, Prussia and Russia. This development was not simply a matter of state-driven policy, the approach usually adopted when considering military change. Instead, the widespread demographic and economic expansion in Europe from the 1740s was also important, producing the resources for greater activity, particularly large numbers of young men, and also more agriculture production.

Greater administrative capability had direct military consequences. This was true of transportation and information. For example, canal construction under Frederick the Great of Prussia was related to the movement of supplies to and from magazines. Earlier, the Austrian conquest of Hungary owed much to the creation of a series of magazines,[3] and, in turn, was followed by immigration and an increase in agricultural production that made it easier to support war.

Mapping was an important aspect of enhanced information provision that interacted with improved military-administrative capability. Large-scale military surveys grew in importance from the eighteenth century. The Austrians, who ruled Sicily between 1720 and 1735, used army engineers to prepare the first detailed map of the island, while Frederick the Great had Silesia mapped. There were also qualitative changes. French military engineers, particularly Pierre Bourçet, improved the mapping of mountains, creating a clearer idea of what the alpine region looked like.[4] Further afield, navies charted coastlines, to the benefit of trade as well as the assertion of power.

Mapping was an aspect of the role of printing. The culture of print was of considerable importance for military development. Printed manuals on gunnery, tactics, drill, fortification and siegecraft spread techniques far more rapidly than word of mouth or manuscript. Manuals also permitted and made possible a degree of standardisation that both helped to increase military effectiveness and was important for the utilisation of military resources. A dynamic relationship between applied science and military practice was fostered. Leonhard Euler, the author of *Neue Grundsatze der Artilleries* (1745), also solved the equations of subsonic ballistic motion in 1753 and summarised some of the results in published tables.

Literacy and printing had other values. They focused discussion of military organisation and methods, and encouraged a sense of system. Translations spread ideas, Guibert's criticism of Frederick the Great's tactics being published not only in French but also in German: *Bemerkungen über die Kriegsverfassung der preussichen Armee. Neue, verbesserte und vermehrte Auflage* (1780). Literacy and printing were also important to command and control and to military administration. It became less difficult to survey the quantity and state of military resources, and military statistics became more accessible. The official report of the Russian Ordnance, *Kniga Pushkarskogo prikaza, za skrepoyu d'yaka Volkova (Book of the Gunners' Chancellery, with the Certification of Secretary Volkov)*(1680) listed all arms and ammunition, as well as current production. Literacy and printing also aided the positive interaction of resources and requirements, of different military branches, in particular localities, and of forces that were in different locations.

At the same time, prior to the French Revolution, knowledge and merit were conceived of in terms that did not challenge seriously the socio-cultural realities of societies organised around the principles of inegalitarianism and inheritance. Larger armies brought more opportunities to nobles, who benefited both from the assumption that they were naturally suited for positions of command and from the fact that in general this was the case. Armies were not forces outside society, but, rather reflections of existing patterns of social and control and influence, and the beliefs that gave cohesion to these patterns. In 1812, 86.5 per cent of Russian officers were noble in their social origins, although, as more generally across the West, noble officership encompassed both the privileged elite and the mass of poor nobles: in 1812, 77 per cent of Russian officers did not own property or serfs.[5] Even in the Dutch Republic, over 60 per cent of colonelcies in the eighteenth century were in the hands of members of the nobility, with the remainder held by members of the urban elite.[6]

The essential stability represented by the symbiotic relationship of military and social hierarchies matched that of the conduct of war, for, although weaponry improved, there was no technological breakthrough and no fundamental alteration in the nature of conflict. The basic possibilities of military action were restricted by technological constraints, particularly in terms of mobility and firepower. There was no rapid technologically driven change comparable to those of the last century or, indeed, 1825–1900. As a result, successful military powers were able to operate on land without altering their economic system or developing a sophisticated industrial capacity. Partly as a consequence, within Europe, there was no qualitative or quantitative superiority sufficient to permit any one state to seize hegemony in what was already a very much a multipolar system. In contrast, the British achieved naval dominance.

The social context of warfare was challenged in the late eighteenth century. British officers in the American War of Independence were surprised by the modest social status of some of the rebel officers. A military ethos centred on the ideal of patriotic citizen-soldiers played a major role in that conflict, as well as in the French Revolutionary wars and the Polish Rising of 1794. In practice, most of the officers in these forces were men of property, but the developments of the period suggested that armies could serve to overthrow established social and political lineaments, rather than to reflect and strengthen them. This toppling or challenging of the *ancien régime* was achieved by the armies of Revolutionary France, and, in so doing, they were certainly revolutionary.

At the same time, the role of contingency was amply displayed by the very different consequences of revolutions in the period 1775–1815. The contrast between success in North America, France and Haiti and failure in the Low Countries, Poland, Ireland and Serbia is readily apparent. International contexts and domestic political tensions could both be crucial, the Poles in 1792–5 falling victim to the former. Success in revolution should rather be traced to particular political circumstances than to unique social structures or discourses of revolution. In Ireland in 1798, French intervention in support of revolution proved far less successful than it had been in North America in 1781, but the insurgents also suffered from the divided nature of Irish political opinion. As with the British in India, this was not the case of an imperial power suppressing a people, but, rather, a more complex and nuanced set of religious, social, political, and economic relations that enabled the government to recruit a significant amount of support, including among Catholics.[7] In Mexico, rebellion against Spanish rule began in 1810, but the royalists were helped by the ethnic character of the rising: largely mestizo, it was seen as a threat to the creoles, as well as to Spanish authority. More generally, the ethnic complexities of Latin America restricted the appeal of radical ideologies and movements.[8]

The international context was important with the successful slave rebellion that occurred on the French Caribbean colony of Saint-Domingue in 1791. Spanish and, even more, British opposition to the French, including, crucially, a blockade of Saint-Domingue's ports in 1803, helped defeat the French efforts, although the resistance of the former slaves was also crucial.[9]

If the transforming character of the development of militaries and, often very differently, the use of force, was one long-term narrative, another was provided by the greater stress on a formalised professionalism based on formal education. This emphasis offered a set of priorities that did not correspond directly to those of the social hierarchy. Efforts to improve officer training in the eighteenth century, mainly in the latter half, were an important innovation and a major pointer to the scientific professionalism of the future. Military education was most developed for artillery, engineering and naval officers; and far less so for infantry and cavalry officers. For example, the Dutch established an artillery school in 1735, and the Spaniards one in 1764. France founded an engineering academy at Mézières in 1748, and Spain a military academy at Zamora in 1790. An artillery school was opened in Naples in 1744, and an engineering academy ten years later, and in 1769 they were amalgamated into the *Reale Accademia Militare*. It taught ballistics, tactics, experimental physics and chemistry to officer cadets, all of whom were supposed to attend classes.[10]

The emphasis on the value of reform and the application of reason to problems which characterised the Enlightenment thought of the period, encouraged a questioning of established precepts and practices and an interest in new ideas. Aside from reform explicitly as change, there was also considerable backing for reform understood as improvement, especially in administrative practices. This helped ensure that governments were better able to exploit their domestic resources, and thus to wage war.

The pace of change and applied knowledge was pushed further during the Napoleonic period, when military schools were founded in Bavaria, Britain and the Netherlands, and an engineering school at Naples. Napoleon founded the École Spéciale Militaire de Saint-Cyr in 1802, to train officers and devoted much time to its syllabus. About 4,000 officers were trained there between 1805 and 1815. Napoleon also reorganised the École Polytechnique in 1804 and the role of mathematics in military education increased.[11] Nevertheless, many officers continued to receive little formal training, in part because high birth, and the qualities this was reputed to betoken and inculcate, continued to be much in demand. Training for ordinary soldiers remained rudimentary, a matter of introducing instruction at the depot and on the march.

The application of knowledge was also a factor in the West's relative military position. Non-Western peoples and states, when confronted by Western aggression, had two advantages: first, at least in South and East Asia and in Africa, superior numbers locally, and, second, greater knowledge of, cultural identity with, and administrative control over local territory. Western success can be explained by a variety of varying combinations of factors, and the way in which the combination held or was held together was in itself important. For example, the British East India Company was, despite its frequent internal disputes, a corporation of seamless continuity and was competing with personalised autocracies which were dependent on strong leadership and which were vulnerable to recurrent succession crises. More generally, the Westerners benefited from the post-feudal, non-personalised nature of their military command systems and command philosophy, especially the emphasis on discipline and the application of reason and science to command problems. Captain Robert Stuart, commander of a sepoy battalion in India in 1773, was convinced that only his firepower-linked definition of discipline would allow his unit to prevail:

> As the superiority of English sepoys over their enemies, as likewise their own safety consist entirely in their steadiness, and attentiveness, to the commands of their officers, it is ordered that no black officer or sepoy pretend to act, or quit his post without positive orders to that purpose from an European officer . . . should any man fire without orders, he is to be put to death upon the spot . . . regularity and obedience to orders are our grand and only superiority.[12]

The application of reason and science was true of weapons development and of tactical theory, which, from the fifteenth-century Renaissance, appears to have been far more highly developed in the West. The number of manuals and speculative works on warfare seems to have been far greater than elsewhere, and, although the process requires explication, this affected aspects of warfare which were hidebound, instinctive and traditional. Moreover, scientific developments were utilised.

Further afield, the attempt to change the Turkish military system pushed by Selim III (r. 1789–1807) indicated another important aspect of change: the internal pressure in non-Western societies to Westernise aspects of their armed forces in order to cope with Western military power and in response to the apparent relative advantage of Western technology and military organisation. Again, the extent of this process should not be exaggerated, because it risks replicating on the international scale the notion of an ideal type and a clear-cut teleology already rejected, in this book, as far as conflict within the West is concerned. Instead, most conflict in which the 'non-West' was concerned did not directly involve Western powers, and this was even more the case prior to the mid-1750s. Conversely, Western powers concentrated, first, on war with each other in Europe itself, and, second, on attacks on other Western colonial possessions. If the French sent a fleet to Sicilian waters in 1674 and the British another in 1718, they were to fight Spanish squadrons, not the Turks, and the goal of opposing the French or Spaniards was true of all major British naval deployments to the Mediterranean from the 1660s until 1807. North African privateering bases represented a different objective, but they were rarely the subject of major and sustained Western attack. The Spaniards made the largest effort, but unsuccessfully so in 1775 and 1784.

In so far as conflict with non-Western powers was concerned, there was no clear pattern of Western success, nor, indeed, consistent pressure. In East Asia, the advance which had taken Western power to the Philippines, Formosa (Taiwan), the Sea of Okhotsk and the Amur Valley by the mid-seventeenth century was reversed, the Dutch being driven from Formosa in 1661–2 and the Russians from the Amur Valley in the 1680s. There was to be no resumption of the pace of Western advance in this region, until British amphibious power was exerted against China in the First Opium War of 1839–42, while Russian expansionism in the Amur region resumed in the late 1850s. Further west, Peter the Great was defeated at the Pruth (1711), the Russians abandoned their gains to the south of the Caspian (1732), and the Austrians surrendered Belgrade to the Turks (1739). The extent to which each of these episodes indicated fundamental flaws can be qualified, but there is no doubt that subsequently Western forces were more successful. This, initially, however, can be located within a wider shift which was not simply a matter of Western success. By 1760 in East Asia, and by 1770 in eastern Europe, the land forces of China and Russia respectively were able readily to defeat opponents which had hitherto resisted with greater success and, as a result, from the 1750s, their land frontiers were pushed outward.

Thanks to the western European powers, however, the West added a dimension lacking in East Asia, that of transoceanic expansion. Although there was no real change in the situation between Westerners and Africans, the frontiers of control and settlement in the New World continued to expand, while, in Australia, the arrival of British forces in 1788 was followed by the rapid establishment of a new military and political order. The Aborigines were not in a position to mount sustained resistance in areas where the environment encouraged large numbers of European settlers: they lacked numbers, firepower and large-scale organisation and were being exposed to new diseases.

The varied role of the West can be seen in contrasting non-Western responses. As yet, China and Japan displayed no interest in changing their military systems, and, for them, the situation did not prefigure that in the late nineteenth and, even more, twentieth century. Nevertheless, Selim III's abortive reforms, as well as the turning to Western military advisers by earlier Turkish rulers, and the use of Western officers and weapons by Indian rulers such as the Nawab of Awadh in the 1760s and the Nizam of Hyderabad in the 1780s–1790s, were indications of the broad pressure for change that in part stemmed from the extension of Western military power, with marked shifts in the sense of what was fit for purpose leading to pressure for development. It is in this global context that Western military history in the period 1660–1815 is of most consequence. The technological changes that clearly established Western military superiority, including railways, iron hulls, rifled guns, steamships and the telegraph did not occur until after 1815, but the Western powers were already dominant at sea. In addition, the fundamental shift within Europe against the Turks occurred from the 1680s onwards. Military strength was central to this rise in Western power, both within and outside Europe, and this strength was to give shape to the nineteenth-century world order.

NOTES

1 INTRODUCTION

1 For scepticism about the notion of technological revolution, A. N. Liaropoulos, 'Revolutions in Warfare: Theoretical Paradigms and Historical Evidence – The Napoleonic and First World War Revolutions in Military Affairs,' *Journal of Military History,* 70 (2006), pp. 363–84.
2 P. W. Schroeder, *The Transformation of European Politics 1763–1848* (Oxford, 1994).
3 C. Ingrao, 'Habsburg Strategy and Geopolitics during the Eighteenth Century', in G. Rothenberg et al. (eds) *East Central European Society and War in the Pre-Revolutionary Eighteenth Century* (New York, 1982), pp. 49–66; J. P. LeDonne, *The Grand Strategy of the Russian Empire, 1650–1831* (Oxford, 2004), on which there is a critical review by R. Frost, *English Historical Review,* 121 (2006), pp. 849–51, and LeDonne, 'Geopolitics, Logistics, and Grain: Russia's Ambitions in the Black Sea Basin, 1737–1834,' *International History Review*, 28 (2006), pp. 1–41.
4 D. J. B. Trim, 'The Continents' Intestines: Estuarine, Riverine and Lacustrine Warfare, c. 1400–1700,' in D. J. B. Trim and M.C. Fissel (eds) *Amphibious Warfare 1000–1700* (Leiden, 2005), pp. 357–419.
5 N. A. M. Rodger, 'Image and Reality in Eighteenth Century Naval Tactics,' *Mariner's Mirror*, 89 (2003), pp. 280–96; S. B. A. Willis, 'Fleet Performance and Capability in the Eighteenth Century Royal Navy,' *War in History*, 11 (2004), pp. 373–92.
6 J. R. Dull, *The French Navy and the Seven Years' War* (Lincoln, Nebr., 2005).
7 BL Add. 61979A fol. 40.
8 R. Jervis, *Perception and Misperception in International Politics* (Princeton, NJ, 1976); A. I. Johnston, *Cultural Realism: Strategic Culture and Grand Strategy in Chinese History* (Princeton, NJ, 1995).
9 Other than for Austria and Russia (see Note 3), there are no equivalents to E. Ringmar, *Identity, Interest and Action: A Cultural Explanation of Sweden's Intervention in the Thirty Years War* (Cambridge, 1996) or G. Parker, *The Grand Strategy of Philip II* (New Haven, Conn., 1998).
10 Memorandum, 'Queries at Turin. 1740,' NA PRO. 30/29/3/1 fol. 16.
11 P. Burke, *The Fabrication of Louis XIV* (New Haven, Conn. 1992).
12 For example, Canale, Sardinian envoy in Vienna, to Charles Emmanuel III, 29 August 1739, AST. Austria 66.
13 Regarding the 1770s, see J. Israel, *The Dutch Republic: Its Rise, Greatness, and Fall, 1477–1806* (Oxford, 1995), p. 1095.
14 G. Plank, *Rebellion and Savagery: The Jacobite Rising of 1745 and the British Empire* (Philadelphia, Pa., 2006).

15 J. B. Hattendorf, *England in the War of the Spanish Succession: A Study of the English View and Conduct of Grand Strategy, 1701–1712* (New York, 1987).
16 J. Luh, '"Strategie und Taktike," im Ancien Régime,' *Militärgeschichtliche Zeitschrift*, 64 (2005), pp. 101–31.
17 George III to William Grenville, Home Secretary, 25 October 1790, BL Add. 58855 fol. 155.

2 CONFLICT BETWEEN WESTERNERS AND NON-WESTERNERS

1 A. J. Smithers, *The Tangier Campaign: The Birth of the British Army* (Stroud, 2003).
2 G. Parker, 'Random Thoughts of a Hedgehog', *Historically Speaking*, 4, no. 4 (April 2003), p. 13.
3 A. Smith, *An Inquiry into the Nature and Causes of the Wealth of Nations* (1776; Oxford, 1976), pp. 689–708.
4 Watson to Robert, 4th Earl of Holdernesse, 7 October 1755, 15 February, 10 March 1756; George Thomas to Mr Thomas, 15 February 1756, BL Eg. 3488 fols 81–2, 140–1, 157–8, 216–17.
5 Cornwallis to Sir Archibald Campbell, 12 April 1788, NA PRO 30/11/159 fols 123–4.
6 BL IO H/Misc/198, pp. 103–4, 112, 36, MSS Eur. Orme OV219 pp. 40–1, 44, 27.
7 BL IO H/Misc/198, pp. 35–6, MSS. Eur. Orme OV219, pp. 26–30.
8 G. B. Malleson, *History of the French in India* (London, 1886), pp. 231–82; V. G. Hatalkar, *Relations between the French and the Marathas, 1668–1815* (Bombay, 1958); G. Bodinier, 'Les Officiers français en Inde de 1750 à 1793', in *Trois siècles de presence française en Inde* (Paris, 1995), pp. 69–89.
9 G. Ágoston, *Guns for the Sultan: Military Power and the Weapons Industry in the Ottoman Empire* (Cambridge, 2005), and 'Behind the Turkish War Machine: Gunpowder Technology and War Industry in the Ottoman Empire, 1450–1700', in B. D. Steele and T. Dorland (eds), *The Heirs of Archimedes. Science and the Art of War through the Age of Enlightenment* (Cambridge, Mass., 2005), pp. 123–5.
10 W. J. Eccles, *Frontenac: The Courtier Governor* (Toronto, 1959), pp. 157–72.
11 J. C. Rule, 'Jerome Phélypeaux, comte de Pontchartrain, and the establishment of Louisiana', in J. F. McDermott (ed.), *Frenchmen and French Ways in the Mississippi Valley* (Urbana, Ill., 1969), pp. 179–97.
12 J. G. Reid, M. Basque, E. Mancke, et al., *The 'Conquest' of Acadia, 1710: Imperial, Colonial, and Aboriginal Constructions* (Toronto, 2004).
13 Bull to General Amherst, 15 April 1761, NA CO 5/61 fol. 277.
14 W. E. Lee, 'Fortify, Fight, or Flee: Tuscarora and Cherokee Defensive Warfare and Military Culture Adaptation', *Journal of Military History*, 68 (2004), p. 770; J. A. Sweet, *Negotiating for Georgia: British-Creek Relations in the Trustee Era, 1733–1752* (Athens, Ga., 2005), p. 144; M. Ward, *Breaking the Backcountry: The Seven Years' War in Virginia and Pennsylvania, 1754–1765* (Pittsburg, Pa., 2003), p. 253.
15 J. Grenier, *The First Way of War: American War Making on the Frontier, 1607–1814* (Cambridge, 2005), pp. 43–52, and *On the Far Reaches of Empire: Anglo-Americans' Fifty Years' War on the Nova Scotia Frontier, 1710–1760* (Vancouver, 2006); M. M. Mintz, *Seeds of Empire: The American Revolutionary Conquest of the Iroquois* (New York, 1999).
16 G. E. Dowd, *War under Heaven: Pontiac, the Indian Nations, and the British Empire* (Baltimore, Md., 2002).
17 M. Giraud, *A History of French Louisiana. V. The Company of the Indies, 1723–1731* (Baton Rouge, La., 1991), pp. 404, 419.
18 P. D. Woods, 'The French and the Natchez Indians in Louisiana, 1700–1731', *Louisiana History*, 19 (1978), pp. 413–35 and *French-Indian Relations on the Southern Frontier, 1699–1762* (Ann Arbor, Mich., 1980).

19 Smith, 'Lectures on Jurisprudence 1762–63', note in *Wealth of Nations,* p. 692, fn. 9.
20 D. Weber, *The Spanish Frontier in North America* (New Haven, Conn., 1992).
21 W. Sunderland, *Taming the Wild Field: Colonization and Empire on the Russian Steppe* (Ithaca, NY, 2004).
22 J. Connor, *The Australian Frontier Wars, 1788–1838* (Sydney, 2002).
23 M. Khodarkovsky, *Where Two Worlds Met: The Russian State and the Kalmyk Nomads, 1600–1771* (Ithaca, NY, 1992), p. 49.
24 V. Lieberman, 'Political Consolidation in Burma under the early Konbaung Dynasty 1750-c. 1820', *Journal of Asian History,* 30 (1996), pp. 161–2.
25 A. Delcourt, *La France et les établissements français au Sénégal entre 1713 et 1763* (Dakar, 1952); J. F. Searing, *West African Slavery and Atlantic Commerce, 1700–1860* (Cambridge, 1993).
26 S. A. Diouf (ed.), *Fighting the Slave Trade: West African Strategies* (Athens, Ohio, 2003).
27 R. G. S. Cooper, 'Culture, Combat, and Colonialism in Eighteenth- and Nineteenth-Century India', *International History Review,* 27 (2005), pp. 548–9.
28 J. Gommans, *Mughal Warfare. Indian Frontiers and High Roads to Empire, 1500–1700* (London, 2002).
29 For example, B. Ganson, *The Guaraní under Spanish Rule in the Río de la Plata* (Palo Alto, Calif., 2003).
30 B. L. Walker, *The Conquest of Ainu Lands: Ecology and Culture in Japanese Expansion, 1590–1800* (Berkeley, Calif., 2001).
31 Ágoston, *Guns for the Sultan*; Cooper, *The Anglo-Maratha Campaigns and the Contest for India: The Struggle for the South Asian Military Economy* (Cambridge, 2004).
32 B. D. Steele, 'Military "Progress" and Newtonian Science in the Age of Enlightenment', in Steele and Dorland (eds), *Heirs of Archimedes,* pp. 381–2.
33 Cooper, *Anglo-Maratha Campaigns.*
34 Cooper, 'Logistics in India, 1757–1857: The Achievement of Arthur Wellesley Reconsidered', in A. J. Guy and P. B. Boyden (eds) *Soldiers of the Raj: The Indian Army, 1600–1947* (London, 1997), pp. 68–77.
35 K. Roy, 'Military Synthesis in South Asia: Armies, Warfare, and Indian Society, c. 1740–1849', *Journal of Military History,* 69 (2005), pp. 689–90.
36 C. Wickremesekera, *'Best Black Troops in the World': British Perceptions and the Making of the Sepoy, 1746–1805* (New Delhi, 2002).
37 BL Add. 13579 fols 7–8; J. Lynn, *Battle: A History of Combat and Culture from Ancient Greece to Modern America* (Boulder, Col., 2003), p. 176.
38 S. Sen, *Empire of Free Trade: The East India Company and the Making of the Colonial Marketplace* (Philadelphia, Pa., 1998).
39 R. J. Barendse, *The Arabian Seas: The Indian Ocean World of the Seventeenth Century* (Armonk, NY, 2002).
40 Barendse, 'Trade and State in the Arabian Seas: A Survey from the Fifteenth to the Eighteenth Century', *Journal of World History,* 11 (2000), pp. 210–11.
41 J. P. LeDonne, 'Geopolitics, Logistics, and Grain: Russia's Ambitions in the Black Sea Basin, 1737–1834', *International History Review,* 28 (2006), pp. 22–26.
42 G. J. Ames, 'Colbert's Grand Indian Ocean Fleet of 1670', *Mariner's Mirror,* 76 (1990), pp. 230–1.
43 M. Sehgal, 'Themes in Late Eighteenth-Century South Asian Military History, 1783–1806' (M. Phil., University of Delhi, 2005).

3 THE NATURE OF CONFLICT

1 R. Muir, *Salamanca 1812* (New Haven, Conn., 2001).
2 B. Capp, *Cromwell's Navy: The Fleet and the English Revolution* (Oxford, 1989); I. Gentles, *The New Model Army in England, Ireland and Scotland, 1645–1653* (Oxford, 1992).

3 A. Forrest, 'Enlightenment, Science and Army Reform in Eighteenth-Century France', in M. Crook, W. Doyle and A. Forrest (eds), *Enlightenment and Revolution: Essays in Honour of Norman Hampson* (Aldershot, 2004), p. 153.
4 J. Langins, *Conserving the Enlightenment: French Military Engineering from Vauban to the Enlightenment* (Cambridge, Mass., 2004).
5 J. Lynn, *Giant of the Grand Siècle: The French Army, 1610–1715* (Cambridge, 1997), pp. 464–65.
6 G. R. Mork, 'Flint and Steel: A Study in Military Technology and Tactics in 17th-Century Europe', *Smithsonian Journal of History*, 2 (1967), pp. 25–52; J. Lynn, 'Forging the Western army in Seventeenth-Century France', in M. Knox and W. Murray (eds), *The Dynamics of Military Revolution 1300–2050* (Cambridge, 2001), pp. 35–56.
7 P. Lenihan, *1690: The Battle of the Boyne* (Stroud, 2003).
8 W. Maltby, 'Politics, Professionalism and the Evolution of Sailing Ships Tactics', in J. A. Lynn (ed.), *Tools of War: Instruments, Ideas and Institutions of Warfare, 1445–1871* (Chicago, Ill., 1990), pp. 53–73; M. A. J. Palmer, 'The Military Revolution Afloat: The Era of the Anglo-Dutch Wars', *War in History*, 4 (1997), pp. 123–49; R. Harding, *Seapower and Naval Warfare 1650–1830* (Basingstoke, 1999), pp. 73–5.
9 G. J. Bryant, 'Asymmetric Warfare: The British Experience in Eighteenth-Century India', *Journal of Military History*, 68 (2004), pp. 458–59.
10 P. J. Speelman, *Henry Lloyd and the Military Enlightenment of Eighteenth-Century Europe* (Westport, Conn., 2002).
11 J. M. Hill, *Celtic Warfare, 1595–1763* (Edinburgh, 1986).
12 S. Brumwell, *Redcoats: The British Soldier and War in the Americas, 1755–1763* (Cambridge, 2002).
13 Sackville to Holdernesse, 10 September 1758, BL Eg. 3444 fol. 66; Fawcett to James Lister, 5 December 1759, Halifax, Calderdale Archives Department SH:7/FAW/58.
14 A. Starkey, *War in the Age of Enlightenment, 1700–1789* (Westport, Conn., 2003), pp. 105–31; D. Grebs, 'The Making of Prisoners of War: Rituals of Surrender in the American War of Independence, 1776–83', *Militärgeschichtliche Zeitschrift*, 64 (2005), pp. 1–29.
15 B. R. Kroener, 'Antichrist, Archenemy, Disturber of the Peace-Forms and Means of Violent Conflict in the Early Middle Ages', in H.-H. Kortüm (ed.), *Transcultural Wars from the Middle Ages to the 21st Century* (Berlin, 2006), p. 78; A. Starkey, 'War and Culture, A Case Study: The Enlightenment and the Conduct of the British Army in America, 1755–81', *War and Society*, 8 (1990), p. 22.
16 Lieutenant Colonel Charles Russell to wife, 5 July 1743, BL Add. 69382.
17 Richard to Jeremy Browne, 14 August 1759, BL RP 3284.
18 Fawcett to Lister, 24 October 1760, Halifax SH:7/FAW/60.
19 M.L. Woodward, 'The Spanish Army and the Loss of America, 1810–24', C. I. Archer, 'The Army of New Spain and the Wars of Independence, 1790–1821', *Hispanic American Historical Review*, 48 (1968), pp. 586–607, 61 (1981), pp. 705–14.
20 Cambridge, University Library, Add. 6570.
21 Ligonier to William, Duke of Cumberland, 24 July 1746, Windsor Castle, Royal Archives, Cumberland Papers 17/252; Martin to 2nd Duke of Richmond, 19 August 1748, Goodwood MSS 107 no. 685.
22 Fawcett to Lister, 24 October 1760, Halifax SH:7/FAW/60; Townshend to 3rd Earl of Bute, 17 September 1761, Mount Stuart, Bute papers 7/23.
23 J. Brewer, *The Sinews of Power: War, Money and the English State, 1688–1783* (London, 1989); J. S. Wheeler, *War and the Military Revolution in Seventeenth-Century England* (Stroud, 1999).
24 J. Childs, 'The English Brigade in Portugal, 1662–68', *Journal of the Society for Army Historical Research*, 53 (1975), pp. 135–47.

25 For only 30,000 effectives in a Sardinian army of 52,000, report by Sardini, envoy of the republic of Lucca, 1 May 1746, Archivio di Stato di Lucca, Anziani al tempo della Libertà, vol. 634.
26 HL Lo. 10125.
27 HL Lo. 8607, 8604, 8608.
28 HL Lo. 10112, 8618.
29 G. Perjés, 'Army Provisioning, Logistics and Strategy in the Second Half of the 17th Century', *Acta Historica Academiae Scientiarum Hungaricae*, 16 (1970), pp. 35–6.
30 Canale, Sardinian envoy in Vienna, to Charles Emmanuel III, 1 October 1740, AST Vienna 67.
31 Oxford, Bodleian Library, MS Eng. Hist. C314 fols 46, 51.
32 B. P. Hughes, *Firepower: Weapons' Effectiveness on the Battlefield, 1680–1850* (London, 1974).
33 D. Showalter, 'Tactics and Recruitment in Eighteenth-Century Prussia', *Studies in History and Politics*, 3 (1983–4), pp. 15–41.
34 BL Add. 61667.
35 Earl of Bristol to Henry Fox, 22 September 1756, NA SP 92/64.
36 A. Forrest, *Napoleon's Men: The Soldiers of the Revolution and Empire* (London, 2002).
37 M. N. McConnell, *Army and Empire. British Soldiers on the American Frontier, 1758–1775* (Lincoln, Nebr., 2005).
38 G. Parker (ed.) *The Cambridge History of Warfare* (Cambridge, 2005), p. 108.

4 WARFARE, 1660–88

1 G. Satterfield, *Princes, Posts and Partisans: The Army of Louis XIV and Partisan Warfare in the Netherlands, 1673–1678* (Leiden, 2003); J. Luh, '"Strategie und Taktik" im Ancien Régime', *Militärgeschichtliche Zeitschrift*, 64 (2005), pp. 117–23; J. Ewald, *Treatise on Partisan Warfare*, edited by R. A. Selig and D. C. Skaggs (Westport, Conn., 1991).
2 L. White, 'Actitudes civiles hacia la Guerra en Extremadura (1640–68)', *Revista de Estudios Extremenos* 43 (1987), pp. 487–501, and 'Estrategia geográfica y fracaso en la reconquista de Portugal por la Monarquía Hispánica, 1640–68', *Studia Histórica. Historia Moderna*, 25 (2003), pp. 59–91.
3 J. Black, *The Age of Total War, 1860–1945* (Westport, Conn., 2006), pp. 47–50.
4 C. Storrs, 'The Army of Lombardy and the Resilience of Spanish Power in Italy in the Reign of Carlos II (1665–1700)' Part I, *War in History*, 4 (1997), pp. 371–97, Part II, *War in History*, 5 (1998), pp. 1–22.
5 G. Hanlon, *The Twilight of a Military Tradition: Italian Aristocrats and European Conflicts, 1560–1800* (London, 1998).
6 P. Wilson, 'War in Early Modern German History', *German History*, 19 (2001), pp. 419–38.
7 J. Lynn, *Giant of the Grand Siècle: The French Army, 1610–1715* (Cambridge, 1997).
8 G. Rowlands, 'Louis XIV, Aristocratic Power and the Elite Units of the French Army', *French History*, 13 (1999), pp. 303–31, 'Louis XIV, Vittorio Amedeo II and French Military Failure in Italy, 1689–96', *English Historical Review*, 115 (2000), pp. 534–69, and *The Dynastic State and the Army under Louis XIV: Royal Service and Private Interest, 1661–1701* (Cambridge, 2002).
9 D. Parrott, *Richelieu's Army: War, Government and Society in France, 1624–1642* (Cambridge, 2002).
10 J. Black, *Kings, Nobles and Commoners: States and Societies in Early Modern Europe. A Revisionist History* (London, 2004).
11 A. James, *The Navy and Government in Early Modern France, 1572–1661* (Woodbridge, 2004).
12 G. Pagès, *Le Grand Électeur et Louis XIV 1660–1688* (Paris, 1905), p. 16.

13 J. Black, *European Warfare, 1494–1660* (London, 2002).
14 L. White, 'Guerra y revolución militar en la Ibéria del siglo XVII', *Manuscrits*, 21 (2003), pp. 63–93.
15 George III to Henry Dundas, Secretary for War, 22 February 1800, BL Add. 40100 fol. 245, cf. 25 July fol. 257.
16 Memoranda on state of Hanoverian army, early 1715, AE CP Brunswick-Hanover 45 fols 4–5.
17 P. Bamford, *Fighting Ships and Prisons: The Mediterranean Galleys of France in the Age of Louis XIV* (Minneapolis, 1973), p. 23.
18 J. Lynn, *Giant of the Grand Siècle: The French Army, 1610–1715* (Cambridge, 1997), pp. 41–64.
19 D. Parrott, 'Cultures of Combat in the *Ancien Régime*: Linear Warfare, Noble Values, and Entrepreneurship', *International History Review*, 27 (2005), pp. 518–33.
20 P. Sonnino, *Louis XIV and the Origins of the Dutch War* (Cambridge, 1988).
21 D. McKay, *The Great Elector* (London, 2001).
22 J. Cornette, *Le Roi de Guerre* (Paris, 1993).
23 R. D. Martin, 'The Marquis of Chamlay' (Ph.D. University of California, Santa Barbara, 1972), pp. 146–55.
24 A. Arkayin, 'The Second Siege of Vienna (1683) and its Consequences', *Revue Internationale d'Histoire Militaire*, 46 (1980), pp. 114–15.
25 J. Stoye, *Marsigli's Europe 1680–1730: The Life and Times of Luigi Ferdinando Marsigli, Soldier and Virtuoso* (New Haven, Conn., 1994), pp. 37–8.
26 C. R. Boxer, 'The Siege of Fort Zeelandia and the capture of Formosa from the Dutch, 1661–62', *Transactions and Proceedings of the Japan Society of London*, 24 (1926–7), pp. 16–47.
27 J. Axtell, *Beyond 1492: Encounters in Colonial North America* (New York, 1992), p. 239.

5 WARFARE, 1689–1721

1 P. C. Perdue, *China Marches West: The Qing Conquest of Central Eurasia* (Cambridge, Mass., 2005).
2 R. Hellie, 'The Petrine Army: Continuity, Change, and Impact', *Canadian-American Slavic Studies*, 8 (1974), p. 253.
3 R. I. Frost, *The Northern Wars 1558–1721* (Harlow, 2000), pp. 268–71.
4 M. C. Paul, 'The Military Revolution in Russia, 1550–1682', *Journal of Military History*, 68 (2004), pp. 42–5.
5 Louis to Verjus, Envoy at the Imperial Diet, 17, 24 December 1688, AE CP Allemagne 323 fols 176, 193.
6 *1691: Le Siège de Mons par Louis XIV* (Brussels, 1991); P. Burke, *The Fabrication of Louis XIV* (New Haven, Conn., 1992), pp. 71–83, 94–102, 110–12.
7 J. Stapleton, 'Grand Pensionary at War: Antonie Heinsius and the Nine Years' War, 1689–97', in J. A. F. de Jongste and A. J. Veenendaal (eds), *Antonie Heinsius and the Dutch Republic 1688–1720: Politics, War and Finance* (The Hague, 2002), pp. 207–8.
8 C. Storrs, 'The Army of Lombardy and the Resilience of Spanish Power in Italy in the Reign of Carlos II, 1665–1700, part 1'; *War in History*, 4 (1997), p. 345.
9 J. Lynn, *The Wars of Louis XIV 1667–1714* (Harlow, 1999), p. 264.
10 J. Ostwald, 'The "Decisive" Battle of Ramillies, 1706: Prerequisites for Decisiveness in Early Modern Warfare', *Journal of Military History*, 64 (2000), pp. 668–77.
11 D. Szechi, *1715: The Great Jacobite Rebellion* (New Haven, Conn., 2006).
12 L. and M. Frey, *Frederick I, the Man and His Times* (Boulder, Col., 1984).
13 Perrone, envoy in Paris, to Victor Amadeus II of Savoy-Piedmont, 11 February, 2 March 1716, in A. Mannon, E. Vayra and E. Ferrero (eds), *Relazioni diplomatiche della*

monarchia di Savoia dalla prima alla seconda restaurazione 1559–1814: Francia . . . 1713–1719 (3 vols, Turin, 1886–91), Vol. II, pp. 108–9, 120.
14 J. Pritchard, *In Search of Empire: The French in the Americas, 1670–1730* (Cambridge, 2003).
15 S. J. Oatis, *A Colonial Complex: South Carolina's Frontiers in the Era of the Yamasee War 1680–1730* (Lincoln, Nebr., 2004).

6 WARFARE, 1722–55

1 The best introduction is D. Showalter, *The Wars of Frederick the Great* (Harlow, 1995).
2 J. Chagniot, *Le Chevalier de Folard, la strategie de l'incertitude* (Paris, 1973).
3 Guy Dickens to Horatio Walpole, 18 September 1736, NA SP 90/41.
4 D. A. Neill, 'Ancestral Voices: The Influence of the Ancients on the Military Thought of the Seventeenth and Eighteenth Centuries,' *Journal of Military History*, 62 (1998), pp. 512–14.
5 M. A. Palmer, '"The Soul's Right Hand": Command and Control in the Age of Fighting Sail, 1652–1827', *Journal of Military History*, 61 (1997), p. 693.
6 E. Lund, *War for the Every Day: Generals, Knowledge, and Warfare in Early Modern Europe, 1680–1740* (Westport, Conn., 1999).
7 Townshend to Robert Walpole, 16 October 1723, NA SP 43/5 fol. 135.
8 K. A. Roider, *The Reluctant Ally: Austria's Policy in the Austro-Turkish War 1737–39* (Baton Rouge, La., 1972) and *Austria's Eastern Question 1700–1790* (Princeton, NJ, 1982).
9 C. Duffy, *The '45. Bonnie Prince Charlie and the Untold Story of the Jacobite Rising* (London, 2003).
10 J. C. Allmager-Beck, 'The Establishment of the Theresan Military Academy in Wiener Neustadt,' in B. K. Király, G. E. Rothenberg and P. F. Sugar (eds), *Essays on Pre-Revolutionary Eighteenth Century East Central European Society and War* (New York, 1982); Barker, *Army, Aristocracy, Monarchy*, pp. 143–4; Exhibition catalogue, *Charles-Alexandre de Lorraine* (Bilzen-Rijkhoven, 1987), pp. 164–5; M. Hochedlinger, '"Bella gerant alii . . . "?: On the State of Early Modern Military History in Austria', *Austrian History Yearbook*, 30 (1999), pp. 237–77.
11 J. P. Merino Navarro, La *Armada Española en el Siglo XVIII* (Madrid, 1981), pp. 51–3; J. Habron, *Trafalgar and the Spanish Navy: The Spanish Experience of Sea Power* (London, 2004), pp. 35–9.

7 WARFARE, 1756–74

1 J. Luh, *Ancien Régime Warfare and the Military Revolution: A Study* (Groningen, 2000), pp. 175, 178.
2 C. Duffy, *Prussia's Glory: Rossbach and Leuthen* (Chicago, Ill., 2003).
3 J. Keep, 'Feeding the Troops: Russian Army Supply Policies during the Seven Years War', *Canadian Slavonic Papers,* 29 (1987), pp. 270, 272.
4 C. Duffy, *The Austrian Army in the Seven Years' War I. Instrument of War* (Rosemont, Ill., 2000).
5 J. A. Gierowski, 'The Polish-Lithuanian Armies in the Confederations and Insurrections of the Eighteenth Centuries', in G. Rothenberg, B. K. Kiraly and P. F. Sugar (eds) *East-Central European Society and War in the Pre-Revolutionary Eighteenth Century* (Boulder, Col., 1982), pp. 233–6.
6 B. W. Menning, 'Russian Military Innovation in the Second Half of the Eighteenth Century', *War and Society*, 2 (1984), p. 37.
7 T. E. Hall, *France and the Eighteenth-Century Corsican Question* (New York, 1971), pp. 187–204.

NOTES

8 WARFARE, 1775–91

1 P. Paret, 'The Relationship between the Revolutionary War and European Military Thought and Practice in the Second Half of the Eighteenth Century', in D. Higginbotham (ed.) *Reconsiderations on the Revolutionary War* (Westport, Conn., 1978), pp. 155–6; W. Bodle, *Valley Forge Winter: Civilians and Soldiers in War* (University Park, Pa., 2002); D. H. Fischer, *Washington's Crossing* (New York, 2004).
2 C. Cox, *A Proper Sense of Honor: Service and Sacrifice in George Washington's Army* (Chapel Hill, NC, 2004).
3 T. E. Chávez, *Spain and the Independence of the United States* (Albuquerque, N. Mex., 2002).
4 D. Hohrath, *Ferdinand Friedrich Nicolai und die militärische Aufklärung im 18. Jahrhundert* (Stuttgart, 1989); P. J. Speelman (ed.) *War, Society and Enlightenment: The Works of General Lloyd* (Leiden, 2005).
5 J. Ewald, *Treatise on Partisan Warfare*, introduced by R. A. Selig and D. C. Skaggs (Westport, Conn., 1991).
6 R. Snowman (ed.) *The Papers of General Nathanael Greene*, Vol. II (Chapel Hill, NC, 1980), p. 235.
7 Jenkinson to his father, Charles, Lord Hawkesbury, 25 July 1792, Bodl. BB 37 fol. 62; Chauvelin to Lebrun, French Foreign Minister, 9 October 1792, AE CP Angleterre 582 fol. 318.
8 C. Duffy, *The Army of Frederick the Great* (London, 1974), pp. 155–6.
9 H. Kleinschmidt, 'Using the Gun: Manual Drill and the Proliferation of Portable Firearms', *Journal of Military History*, 63 (1999), p. 617.
10 R. Quimby, *The Background of Napoleonic Warfare: The Theory of Military Tactics in Eighteenth-Century France* (New York, 1957); S. T. Ross, 'The Development of the Combat Division in Eighteenth-Century French Armies', *French Historical Studies*, 1 (1965), pp. 85–6.
11 M. Hochedlinger, *Austria's Wars of Emergence 1683–1797* (London, 2003), p. 304.
12 H. M. Scott, 'Aping the Great Powers: Frederick the Great and the Defence of Prussia's International Position, 1763–86', *German History*, 12 (1994), pp. 286–307.
13 D. Hume, *The History of England from the Invasion of Julius Caesar to the Revolution in 1688* (6 vols, 1778, London) Vol. II, p. 230.
14 R. Stone, *The Genesis of the French Revolution: A Global-Historical Interpretation* (Cambridge, 1994); O. T. Murphy, *The Diplomatic Retreat of France and Public Opinion on the Eve of the French Revolution, 1783–1789* (Washington, DC, 1998).
15 A. Balisch, 'Infantry Battlefield Tactics in the Seventeenth and Eighteenth Centuries on the European and Turkish Theatres of War: The Austrian Response to Different Conditions', *Studies in History and Politics*, 3 (1983–4), pp. 52–9.
16 J. Glete, 'The Foreign Policy of Gustavus III and the Navy as an Instrument of that Policy', in *The War of King Gustavus III and Naval Battles of Routsinsalmi* (Kotka, 1993), pp. 5–12.

9 WARFARE IN THE FRENCH REVOLUTIONARY AND NAPOLEONIC ERA, 1792–1815

1 W. Cobbett (ed.) *Parliamentary History of England*, vol. 33 (London, 1818) cols 226–7.
2 D. Moran and A. Waldron (eds) *The People in Arms: Military Myth and National Mobilization since the French Revolution* (Cambridge, 2003).
3 S. P. Mackenzie (ed.) *Revolutionary Armies in the Modern Era: A Revisionist Approach* (London, 1997).
4 J. Lynn, *The Bayonets of the Republic: Motivation and Tactics in the Army of Revolutionary France* (Urbana, Ill., 1984).

5 T. Tackett, 'Conspiracy Obsession in a Time of Revolution: French Elites and the Origins of the Terror, 1789–92', *American Historical Review*, 105 (2000), pp. 691–713.
6 J. Lukowski, *Liberty's Folly: The Polish-Lithuanian Commonwealth in the Eighteenth Century, 1697–1795* (Harlow, 1991).
7 R. Blaufarb, *The French Army, 1750–1820: Careers, Talent, Merit* (Manchester, 2003).
8 H. G. Brown, *War, Revolution and the Bureaucratic State: Politics and the Army Administration in France, 1791–1799* (Oxford, 1995).
9 K. Alder, *Engineering the Revolution: Arms and Enlightenment in France, 1763–1815* (Princeton, NJ, 1997).
10 P. Griffith, *The Art of War of Revolutionary France, 1789–1802* (London, 1998), p. 282.
11 D. Showalter, 'Hubertusberg to Auerstädt: The Prussian Army in Decline?', *German History*, 12 (1994), pp. 308–33.
12 R. Holmes, *Redcoat: The British Soldier in the Age of Horse and Musket* (London, 2002).
13 G. Rothenburg, *Napoleon's Great Adversaries: The Archduke Charles and the Austrian Army, 1792–1814* (London, 1982).
14 R. Blaufarb, '"The Military Lays Open the Civil, and the Civil Betrays the Military Anarchy": The Structure of Civil-Military Relations, 1789–91', *Consortium on Revolutionary Europe: Selected Papers, 1996* (Tallahassee, Fla., 1996), pp. 71–7.
15 J. North, 'General Hoche and Counterinsurgency', *Journal of Military History*, 67 (2003), p. 532.
16 L. W. Eysturlid, *The Formative Influencers, Theories, and Campaigns of the Archduke Carl of Austria* (Westport, Conn., 2000).
17 R. E. Parrish, 'Jacques Étienne Macdonald: Military Administrator in the Roman Republic, 1798', *Consortium on Revolutionary Europe: Selected Papers, 1996* (Tallahassee, Fla., 1996), p. 201.
18 P. Longworth, *The Art of Victory: The Life and Achievements of Generalissimo Suvorov, 1729–1800* (London, 1965).
19 E. Dague, 'Building the Ministry of War: The Consular Period, 1799–1801', *Consortium on Revolutionary Europe: Selected Papers, 2000* (Tallahassee, Fla., 2000), pp. 129–38.
20 A. Forrest, *Napoleon's Men: The Soldiers of the Revolution and Empire* (London, 2002).
21 R. M. Epstein, 'Patterns of Change and Continuity in Nineteenth-Century Warfare', *Journal of Military History*, 56 (1992), pp. 376–7.
22 B. McConachy, 'The Roots of Artillery Doctrine: Napoleonic Artillery Tactics Reconsidered', *Journal of Military History*, 65 (2001), p. 632.
23 O. Connelly, *Blundering to Glory: Napoleon's Military Campaigns* (2nd edn, Wilmington, Del., 1999).
24 F. C. Schneid, *Napoleon's Conquest of Europe: The War of the Third Coalition* (Westport, Conn., 2005), pp. 106–10.
25 F. C. Schneid, *Napoleon's Italian Campaigns, 1805–1815* (Westport, Conn., 2002).
26 M. Finley, *The Most Monstrous of Wars: The Napoleonic Guerrilla War in Southern Italy, 1806–1811* (Columbia, SC, 1994).
27 D. Stefanović, 'Seeing the Albanians through Serbian Eyes: The Inventors of the Tradition of Intolerance and Their Critics, 1804–1939', *European History Quarterly*, 35 (2005), pp. 465–7.
28 J. L. Tone, *The Fatal Knot: The Guerilla War in Navarre and the Defeat of Napoleon in Spain* (Chapel Hill, NC, 1994).
29 R. M. Epstein, *Napoleon's Last Victory and the Emergence of Modern War* (Lawrence, Kans., 1994).
30 J. Gill, '"Those Miserable Tyrolians": Battle Tactics and Alliance Politics in 1809', *Consortium on Revolutionary Europe: Selected Papers, 2001* (Tallahassee, Fla., 2003), p. 253.
31 W. Shanahan, *Prussian Military Reforms, 1788–1813* (New York, 1945); P. Paret, *Yorck and the Era of Prussian Reform, 1807–1815* (Princeton, NJ, 1966); D. Showalter,

'Manifestation of Reform: The Rearmament of the Prussian Infantry, 1806–13', *Journal of Modern History*, 44 (1972), pp. 364–80; R. E. Franck, 'Innovation and the Technology of Conflict during the Napoleonic Revolution in Military Affairs', *Conflict Management and Peace Science*, 21 (2004), pp. 69–84.

32 C. J. Esdaile, *The Spanish Army in the Peninsular War* (Manchester, 1988).

33 H. T. Parker, 'Why Did Napoleon Invade Russia? A Study in Motivation and the Interrelations of Personality and Social Structure', and P. W. Schroeder, 'Napoleon's Foreign Policy: A Criminal Enterprise', *Journal of Military History*, 54 (1990), pp. 131–46, 147–62.

34 F. C. Schneid, 'The Dynamics of Defeat: French Army Leadership, December 1812–March 1813', *Journal of Military History*, 63 (1999), pp. 7–28.

35 J. Gill, 'Impossible Numbers: Solving Rear Area Security Problems in 1809', *Consortium on Revolutionary Europe: Selected Papers, 2000* (Tallahassee, Fla., 2000), pp. 212–22.

36 V. Leggiere, 'From Berlin to Leipzig: Napoleon's Gamble in North Germany, 1813', *Journal of Military History*, 67 (2003), pp. 39–84.

37 D. Smith, *1813 Leipzig: Napoleon and the Battle of the Nations* (Mechanicsburg, Pa., 2001).

38 A. Roberts, *Napoleon and Wellington* (London, 2001).

39 J. Lynn, 'Napoleonic Warfare, 1805–7: Model or Special Case', *Consortium on Revolutionary Europe: Selected Papers, 1998* (Tallahassee, Fla., 1998), pp. 97–105.

10 NAVAL POWER

1 I am most grateful to Jan Glete, Richard Harding and Sam Willis for their comments on an earlier draft of this chapter.

2 P. C. Perdue, *China Marches West: The Qing Conquest of Central Eurasia* (Cambridge, Mass., 2005).

3 For European equivalents, D. J. B. Trim, 'Medieval and Early-Modern Inshore, Estuarine, Riverine and Lacustrine Warfare', in D. J. B. Trim and M. C. Fissel (eds) *Amphibious Warfare 1000–1700: Commerce, State Formation and European Expansion* (Leiden, 2006), pp. 357–420.

4 R. Tregaksis, *The Warrior King: Hawaii's Kamehameha the Great* (New York, 1973).

5 See also H. Moyse-Bartlett, *The Pirates of Trucial Oman* (London, 1966) and L. R. Wright, 'Piracy in the Southeast Asian Archipelago', *Journal of Oriental Studies*, 14 (1976), pp. 23–33; B. Sandin, *The Sea Dayaks of Borneo: Before White Rajah Rule* (London, 1967).

6 C. O. Philip, *Robert Fulton* (New York, 1985), p. 302.

7 P. P. Bernard, 'How Not to Invent the Steamship', *East European Quarterly*, 14 (1980), pp. 1–8.

8 L. Levathes, *When China Ruled the Seas: The Treasure Fleet of the Dragon Throne 1405–1433* (New York, 1995); S. Turnbull, *Samurai Invasion* (London, 2002); K. M. Swope, 'Crouching Tigers, Secret Weapons: Military Technology Employed During the Sino-Japanese-Korean War, 1592–98', *Journal of Military History*, 69 (2005), pp. 31–4.

9 J. Glete, *Warfare at Sea, 1500–1650* (London, 2000) and *War and the State in Early Modern Europe* (London, 2002); N. A. M. Rodger, *The Safeguard of the Sea* (London, 1997) and *The Command of the Ocean* (London, 2004).

10 R. Harding, *Seapower and Naval Warfare 1650–1830* (London, 1999), p. 205.

11 S. Chaudhury and M. Morineau (eds) *Merchants, Companies and Trade: Europe and Asia in the Early Modern Era* (Cambridge, 1999).

12 T. Andrade, 'The Company's Chinese Pirates: How the Dutch East India Company Tried to Lead a Coalition of Pirates to War against China, 1621–62', *Journal of World History*, 15 (2005), pp. 442–4.

13 D. Crecelius, 'Egypt in the Eighteenth Century', in M. W. Daly (ed.) *The Cambridge History of Egypt. II. Modern Egypt* (Cambridge, 1998), p. 60; J. Hathaway, *The Politics of Households in Ottoman Egypt: The Rise of the Qazdaglis* (Cambridge, 1997).
14 C. Totman, *Early Modern Japan* (Berkeley, Calif., 1994).
15 J.A. Millward, *Beyond the Pass: Economy, Ethnicity and Empire in Qing Central Asia, 1759–1864* (Stanford, Calif., 1998).
16 M. Malgonkar, *Kanhoji Angrey, Maratha Admiral* (Bombay, 1959).
17 L. Lockhart, 'Nadir Shah's Campaigns in Oman, 1734–44', *Bulletin of the School of Oriental and African Studies,* 8 (1935–7), pp. 157–73.
18 B. Vale, *A War Betwixt Englishmen: Brazil Against Argentina on the River Plate, 1825–1830* (London, 2000).
19 A. DeConde, *The Quasi-War: The Politics and Diplomacy of the Undeclared War with France, 1797–1801* (New York, 1966).
20 S. C. Tucker, *The Jeffersonian Gunboat Navy* (Columbia, SC, 1993); C. L. Symonds, *Navalists and Antinavalists: The Naval Policy Debate in the United States, 1785–1827* (Newark, Del., 1980).
21 A. Deshpande, 'Limitations of Military Technology: Naval Warfare on the West Coast, 1650–1800', *Economic and Political Weekly,* 25 (1992), pp. 902–3.
22 J. C. Beaglehole, *The Exploration of the Pacific* (3rd edn, Stanford, Calif., 1966).
23 G. V. Scammell, *The World Encompassed: The First European Maritime Empires, c. 800–1650* (London, 1981).
24 J. P. Merino Navarro, *La Armada Española en el Siglo XVIII* (Madrid, 1981), p. 168.
25 J. Glete, *Navies and Nations: Warships, Navies and State Building in Europe and America, 1500–1860* (2 vols, Stockholm, 1993) Vol. I, p. 313.
26 R. Morriss, *The Royal Dockyards during the Revolutionary and Napoleonic Wars* (Leicester, 1983); C. Wilkinson, *The British Navy and the State in the Eighteenth Century* (Woodbridge, 2004).
27 M. Duffy, 'The Establishment of the Western Squadron as the Linchpin of British Naval Strategy', in M. Duffy (ed.) *Parameters of British naval power 1650–1850* (Exeter, 1992), pp. 61–2 and 'The Creation of Plymouth Dockyard and Its Impact on Naval Strategy', in *Guerres Maritimes 1688–1713* (Vincennes, 1990), pp. 245–74.
28 J. E. Talbott, *The Pen and Ink Sailor: Charles Middleton and the King's Navy, 1778–1813* (London, 1998); C. Wilkinson, *The British Navy and the State in the Eighteenth Century* (Woodbridge, 2004).
29 J. R. Dull, *The French Navy and the Seven Years' War* (Lincoln, Nebr., 2005).
30 J. M. Stapleton, 'The Blue-Water Dimension of King William's War: Amphibious Operations and Allied Strategy during the Nine Years' War, 1688–97' in Trim and Fissel (eds), *Amphibious Warfare*, p. 348.
31 Rockingham to Earl of Hardwicke, *c.* Ap. 1781, Sheffield, City Archive, Wentworth Woodhouse MSS R1-1962.
32 J. E. Talbott, 'Copper, Salt, and the Worm', *Naval History,* 3 (1989), p. 53, and 'The Rise and Fall of the Carronade', *History Today,* 39, 8 (1989), pp. 24–30; R. J. W. Knight, 'The Royal Navy's Recovery After the Early Phase of the American Revolutionary War', in G. J. Andreopoulos and H. E. Selesky (eds) *The Aftermath of Defeat: Societies, Armed Forces, and the Challenge of Recovery* (New Haven, Conn., 1994), pp. 10–25; R. Cock, 'The Finest Invention in the World: The Royal Navy's Early Trials of Copper Sheathing, 1708–70', *Mariner's Mirror,* 87 (2001), pp. 446–59.
33 M. Duffy, 'The Gunnery at Trafalgar: Training, Tactics or Temperament?', *Journal for Maritime Research* (August 2005), available online at http://www.jmr.ac.uk.
34 Glete, *Navies,* pp. 402, 405. Glete employs a different system of measurement from that traditionally used and the tonnages are therefore higher by 500–750 tons than normally given.

35 Rodger, *Command of the Ocean*, p. 422.
36 M. Duffy, '". . . All Was Hushed Up": The Hidden Trafalgar', *Mariner's Mirror*, 91 (2005), pp. 216–40.
37 J. Gwyn, *Ashore and Afloat: The British Navy and the Halifax Naval Yard before 1820* (Toronto, 2004).
38 R. Buel Jr., *In Irons: Britain's Naval Supremacy and the American Revolutionary Economy* (New Haven, Conn., 1999).
39 P. Krajeski, 'The Foundation of British Amphibious Warfare Methodology During the Napoleonic Era, 1793–1815', *Consortium on Revolutionary Europe: Selected Papers 1996* (Tallahassee, Fla., 1996), pp. 191–8.
40 D. Syrett, *The Royal Navy in American Waters, 1775–1783* (Aldershot, 1989).
41 N. A. M. Rodger, *The Insatiable Earl: A life of John Montagu, 4th Earl of Sandwich* (London, 1993), pp. 365–77.
42 J. R. Dull, *The French Navy and American Independence: A Study of Arms and Diplomacy, 1774–1787* (Princeton, NJ, 1975).
43 S. Willis, 'The Capability of Sailing Warships, part 2: Manoeuvrability', *Le Marin du Nord/The Northern Mariner*, 14 (2004), pp. 57–68.
44 N. A. M. Rodger, 'Image and Reality in Eighteenth Century Naval Tactics', *Mariner's Mirror*, 89 (2003), pp. 281–6.
45 S. Willis, 'Fleet Performance and Capability in the Eighteenth-Century Royal Navy', *War in History*, 11 (2004), pp. 373–92.
46 For the role of the weather gauge, not least the contrast between theory and practice, S. Willis, 'Capability, Control and Tactics in the Eighteenth-Century Royal Navy' (University of Exeter Ph.D., 2004), pp. 203–28.
47 Blankett to Earl of Shelburne, 29 July 1778, BL, Bowood papers 511 fols 9–11.
48 W. S. Cormack, *Revolution and Political Conflict in the French Navy, 1789–1794* (Cambridge, 1995).
49 George III to William Pitt the Younger, First Lord of the Treasury, 4 March 1797, PRO 30/8/104 fol. 145.
50 D. D. Howard, 'British Seapower and its Influence on the Peninsular War, 1810–18', *Naval War College Review*, 21 (1978), pp. 54–71; C. D. Hall, 'The Royal Navy and the Peninsular War', *Mariner's Mirror*, 79 (1993), pp. 403–18.
51 J. Pritchard, *Anatomy of a Naval Disaster* (Montreal, 1996).
52 C. Crewe, *Yellow Jack and the Worm: British Naval Administration in the West Indies, 1739–1748* (Liverpool, 1993).
53 L. Maloney, 'The War of 1812: What Role for Sea Power?', in K. J. Hagan (ed.) *This People's Navy: The Making of American Sea Power* (New York, 1991), pp. 46–62; R. Morriss, *Cockburn and the British Navy in Transition: Admiral Sir George Cockburn, 1772–1853* (Exeter, 1997), pp. 83–120.
54 C. Ware, 'The Glorious First of June: The British Strategic Perspective', in M. Duffy and R. Morriss (eds) *The Glorious First of June 1794: A Naval Battle and its Aftermath* (Exeter, 2001), pp. 38–40.
55 P. Mackesy, *British Victory in Egypt, 1801: The End of Napoleon's Conquest* (London, 1995).
56 George III to George, 2nd Earl Spencer, 1st Lord of the Admiralty, 17 April 1795, BL Add. 75779.
57 R. Morriss (ed.) *The Channel Fleet and the Blockade of Brest, 1792–1801* (Aldershot, 2001).
58 P. Mackesy, *The War in the Mediterranean, 1803–1810* (London, 1957); P. Krajeski, *In the Shadow of Nelson: The Naval Leadership of Admiral Sir Charles Cotton, 1753–1812* (Westport, Conn., 2000).
59 C. Ware, *The Bomb Vessel: Shore Bombardment Ships of the Age of Sail* (Annapolis, Md., 1994).

NOTES

60 A. Roland, *Underwater Warfare in the Age of Sail* (Bloomington, Ind., 1978); W. S. Hutcheon, *Robert Fulton, Pioneer of Undersea Warfare* (Annapolis, 1981); G. L. Pesce, *La Navigation sous-marine* (Paris, 1906), p. 227.
61 John Robinson, Secretary to the Treasury, to George III, 2 November 1775, BL Add. 37833 fol. 18.
62 J. Pritchard, *Louis XV's Navy 1748–1762: A Study of Organisation and Administration* (Montreal, 1987); W. S. Cormack, *Revolution and Political Conflict in the French Navy, 1789–1794* (Cambridge, 1995).
63 Fulton to William, Lord Grenville, British Prime Minister, 2 September 1806, BL Add., vol. 71593 fol. 134.

11 SOCIAL AND POLITICAL CONTEXTS

1 J. E. Thomson, *Mercenaries, Pirates and Sovereigns. State-Building and Extra-Territorial Violence in Early Modern Europe* (Princeton, NJ, 1994).
2 George to Henry Dundas, 17 July 1791, BL Add. 40100 fol. 4.
3 P. Harsin, *Les Rélations Extérieures de la Principauté de Liège* (Liège, 1927), pp. 229–30.
4 J. R. Fisher, A. J. Kuethe and A. McFarlane (eds) *Reform and Insurrection in Bourbon New Granada and Peru* (Baton Rouge, La., 1990).
5 M. P. McKinley, *Pre-Revolutionary Caracas: Politics, Economy and Society 1777–1811* (Cambridge, 1985).
6 P. Wilson, 'German Women and War, 1500–1800', *War in History*, 3 (1996), p. 147.
7 A. Espino López, 'L'evolució de les forces auxiliars Durant la Guerra de Successió a Catalunya, 1705–14. El cas dels miquelets i dels voluntaris', *Manuscrits*, 52 (2005), pp. 541–56.
8 J. Ruff, *Violence in Early Modern Europe 1500–1800* (Cambridge, 2001).
9 J. Moore, *A View of Society and Manners in France, Switzerland and Germany* (2 vols, London, 1779), Vol. I, p. 383.
10 J. Lynn, *The Wars of Louis XIV 1667–1714* (Harlow, 1999), pp. 22–3.
11 L. Dubois, *Avengers of the New World: The Story of the Haitian Revolution* (Cambridge, Mass., 2004); A. S. Marks, 'The Statue of King George III in New York and the Iconology of Regicide', *The American Art Journal*, 13 (1981), pp. 61–82.
12 P. Sonnino, *Louis XIV and the Origins of the Dutch War* (Cambridge, 1988), p. 64; C. Ingrao, *The Hessian Mercenary State. Ideas, Institutions and Reform under Frederick II 1760–1785* (Cambridge, 1987), p. 129.
13 George III to Lord North, 16 October 1775, J. Fortescue (ed.) *The Correspondence of King George the Third* (6 vols, London, 1927–8), Vol. III, p. 270.
14 George III to Henry Dundas, 9 October 1799, BL Add. 40100 fol. 235.
15 P. Mansel, 'Monarchy, Uniform and the Rise of the Frac 1760–1830', *Past and Present*, 96 (1982), pp. 103–32.
16 R. Browning, 'New Views on the Silesian Wars', *Journal of Military History*, 69 (2005), pp. 532–3.
17 D. Chandler (ed.) *Napoleon's Marshals* (London, 1987), pp. xxxviii–xxxxix.
18 R. Butler, *Choiseul. I. Father and Son 1719–1754* (Oxford, 1980), p. 661.
19 BL Add. 71172 fol. 27.
20 M. Roberts, *Swedish Imperial Experience*, pp. 60–1 and *The Age of Liberty. Sweden 1719–1772* (Cambridge, 1986), p. 75; C. Storrs and H. M. Scott, 'The Military Revolution and the European Nobility, c. 1600–1800', *War in History*, 3 (1996), pp. 18–19.
21 G. Hanlon, *The Twilight of a Military Tradition: Italian Aristocrats and European Conflicts, 1560–1800* (London, 1998).
22 George III to Jenkinson, 29 June 1779, BL Loan 72/1 fol. 13.

NOTES

23 George III to George, 2nd Earl Spencer, First Lord of the Admiralty, 17 March 1795, BL Add. 75779.
24 J. Smith, *The Culture of Merit: Nobility, Royal Service, and the Making of Absolute Monarchy in France, 1600–1789* (Ann Arbor, Mich., 1996).
25 P. Wilson, 'Violence and the Rejection of Authority in Eighteenth-Century Germany: The Case of the Swabian Mutinies in 1757', *German History*, 12 (1994), pp. 1–26.
26 P. Wilson, 'The German "Soldier Trade" of the Seventeenth and Eighteenth Centuries: A Reassessment', *International History Review*, 19 (1996), pp. 757–92.
27 P. Wilson, 'The Politics of Military Recruitment in Eighteenth-Century Germany', *English Historical Review*, 117 (2002), p. 567.
28 W. Fann, 'Peacetime Attrition in the Army of Frederick William I, 1713–40', *Central European History*, 11 (1978), pp. 323–34.
29 D. Showalter, 'The Prussian Military State', in G. Mortimer (ed.) *Early Modern Military History* (Basingstoke, 2004), pp. 123–5.
30 P. Wilson, 'Social Militarisation in Eighteenth-Century Germany', *German History*, 18 (2000), pp. 1–39.
31 Ingrao, *The Hessian Mercenary State*; P. K. Taylor, *Indentured to Liberty: Peasant Life and the Hessian Military State, 1688–1815* (Ithaca, NY, 1994).
32 Jenkinson to Amherst, 24 October 1780, NA WO 34/127 fol. 155.
33 Silhouette to Amelot, French Foreign Minister, 7 September 1741, NA SP 107/49; R. N. Buckley, *Slaves in Red Coats: The British West India Regiments, 1793–1815* (New Haven, Conn., 1979); C. J. Bartlett and G. A. Smith, 'A "Species of Milito-Nautico-Guerilla-Plundering Warfare": Admiral Alexander Cochrane's Naval Campaign against the United States, 1814–15', in J. Flavell and S. Conway (eds) *Britain and America Go to War: The Impact of War and Warfare in Anglo-America, 1754–1815* (Gainesville, Fla., 2004), pp. 187–90.
34 Lynn, *Giant of the Grand Siècle*, pp. 380–93.
35 D. M. Hopkin, *Soldier and Peasant in French Popular Culture, 1766–1870* (Woodbridge, 2003).
36 J. C. Riley, *The Seven Years' War and the Old Regime in France: The Economic and Financial Toll* (Princeton, NJ, 1987), pp. 78–9, 103.
37 A. Corvisier, *L'Armée française de la fin du XVIIe siècle au ministère de Choiseul* (2 vols, Paris, 1964), Vol. I, pp. 220–1, 229.
38 A. Forrest, *Conscripts and Deserters: The Army and French Society during the Revolution and Empire* (London, 1989).
39 P. Cuccia, '"Giornale Della Libertà": The French Occupation and Administration of Mantua 1797–99', *Consortium on Revolutionary Europe: Selected Papers, 2001* (Tallahassee, Fla., 2003), p. 182; F. C. Schneid, *Soldiers of Napoleon's Kingdom of Italy: Army, State and Society, 1800–1814* (Boulder, Col., 1995).
40 I. Woloch, 'Napoleonic Conscription: State Power and Civil Society', *Past and Present*, 111 (1986); S. Woolf, *Napoleon's Integration of Europe* (London, 1991), p. 171.
41 J. P. Bois, 'Les anciens soldats de 1715 à 1815. Problèmes et méthodes', *Revue Historique*, 265 (1981), pp. 81–102, and 'Les Soldats invalids au XVIIIème siècle. Perspectives nouvelles', *Histoire, Economie et Sociale*, 2 (1982), pp. 237–58.
42 M. Leclère, 'Les reformes de Castries', *Revue des Questions Historiques*, 128 (1937), pp. 28–62.
43 C. Duffy, *The Military Life of Frederick the Great* (London, 1986), p. 335.
44 B. Marshke, *Absolutely Pietist: Patronage, Factionalism, and State-Building in the Early Eighteenth-Century Prussian Army Chaplaincy* (Tubingen, 2005).
45 A. Forrest, *Napoleon's Men: The Soldiers of the Revolution and Empire* (London, 2002).
46 A. V. Berkis, *The Reign of Duke James in Courland, 1638–1682* (Lincoln, Nebr., 1960), pp. 140–1.

47 D. D. Horward (ed.) *The French Campaign in Portugal, 1810–1811. An account by Jean Jacques Pelet* (Minneapolis, Minn., 1973), pp. 13, 15; G. Rowlands, 'Louis XIV, Vittorio Amedeo II and French Military Failure in Italy, 1689–96', *English Historical Review*, 115 (2000), p. 543.
48 C. J. Esdaile, *Fighting Napoleon: Guerrillas, Bandits and Adventurers in Spain, 1808–1814* (New Haven, Conn., 2004).
49 A. D. Francis, *The First Peninsular War* (London, 1975), p. 247.
50 P. Joutard, *Les Camisards* (Paris, 1976); L. and M. Frey, *Societies in Upheaval: Insurrections in France, Hungary, and Spain in the Early Eighteenth Century* (London, 1987), pp. 49–56.
51 A. Espino López, 'La Situación Militar en el Reino de Valencia durante la Segunda Germanía (1693)', *Pedralbes*, 24 (2004), pp. 233–84.
52 Nattrass, 'Swiss Civil War of 1712', p. 17, Francis, *Peninsular War*, pp. 371–9.
53 Edinburgh, National Library of Scotland, MS 16630, fol. 66.
54 J. Black, *Culloden and the '45* (Stroud, 1990), p. 178.
55 Ibid.
56 Earl of Ilchester (ed.) *Letters to Henry Fox* (London, 1915), pp. 9, 13–14; NA SP 54/30 fols 12–13; Albemarle Papers, Vol. I, p. 5.
57 BL Add. 23810 fol. 368.
58 J. Basarb, *Pereiaslav 1654: An Historiographical Study* (Edmonton, 1982), p. 71.
59 Reading, Berkshire CRO Trumbull MS 133/11/1; BL Add. 57308 fol. 77.
60 BL Add. 61979A fol. 54; Moore, *A View of Society and Manners*, Vol. I, p. 395.
61 G. Best, *War and Society in Revolutionary Europe, 1770–1870* (London, 1982), pp. 168–83.
62 Dundas, memorandum, – October 1796, BL Add. 59280 fols 189–92.

12 CONCLUSIONS

1 L. Woolf, *Inventing Eastern Europe: The Map of Civilization on the Mind of the Enlightenment* (Stanford, Calif., 1994); M. T. Poe, *'A People Born to Slavery.' Russia in Early Modern European Ethnography, 1475–1748* (Ithaca, NY, 2001).
2 O. Subtelny, *Domination of Eastern Europe: Native Nobilities and Foreign Absolutism 1500–1715* (Gloucester, 1986); P. Longworth, *The Cossacks* (London, 1969); J. E. Thomson, *Mercenaries, Pirates, and Sovereigns: State-Building and Extraterritorial Violence in Early Modern Europe* (Princeton, NJ, 1994).
3 J. Nouzille, 'Charles V de Lorraine, les Habsbourg et la guerre contre les Turcs de 1683 à 1687', in J. P. Bled, E. Faucher and R. Tavenaux (eds) *Les Habsbourg et la Lorraine* (Nancy, 1988), p. 112.
4 J. Pallière, 'Un grand méconnu du XVIIIe siècle: Pierre Bourçet, 1700–1780', *Revue Historique des Armées* (1979), pp. 51–66.
5 A Mikaberidze, *The Russian Officer Corps in the Revolutionary and Napoleonic Wars, 1792–1815* (El Dorado Hills, Calif., 2005).
6 H. L. Zwitzer, *De Mitie van den Staat* (Amsterdam, 1991), p. 241.
7 D. Keogh and N. Furlong (eds) *The Mighty Wave: The 1798 Rebellion in Wexford* (Blackrock, 1996).
8 C. I. Archer, 'Insurrection-Reaction-Revolution-Fragmentation: Reconstructing the Choreography of Meltdown in New Spain during the Independence Era', *Mexican Studies*, 10 (1994), pp. 63–98; V. Guedea, 'The Process of Mexican Independence', *American Historical Review*, 105 (2000), pp. 116–30.
9 C. L. R. James, *The Black Jacobins: Toussaint L'Ouverture and the San Domingo Revolution* (London, 1980); D. P. Geggus, 'The Haitian Revolution', in F. W. Knight and C. A. Palmer (eds) *The Modern Caribbean* (Chapel Hill, NC, 1989), pp. 21–50.

10 G. Hanlon, *The Twilight of a Military Tradition: Italian Aristocrats and European Conflicts, 1560–1800* (London, 1998), p. 346.
11 J. Langins, 'The École Polytechnique and the French Revolution: Merit, Militarization, and Mathematics', *Llull*, 13 (1990), pp. 91–105.
12 Stuart, orders, BL Add. 29198 fols 120, 123.

SELECT BIBLIOGRAPHY

1 GENERAL

M. S. Anderson, *War and Society in Europe of the old regime, 1618–1789* (London, 1988).
G. Best, *War and Society in Revolutionary Europe, 1770–1870* (London, 1982).
J. Childs, *Armies and Warfare in Europe, 1648–1789* (Manchester, 1982).
G. N. Clark, *War and Society in the Seventeenth Century* (Cambridge, 1958).
A. Corvisier, *Armies and Societies in Europe, 1494–1789* (Bloomington, 1979).
G. Hanlon, *The Twilight of a Military Tradition. Italian Aristocrats and European Conflicts, 1560–1800* (London, 1998).
J. M. Hill, *Celtic Warfare 1595–1763* (London, 1986).
G. Mortimer (ed.) *Early Modern Military History, 1450–1815* (Basingstoke, 2004).
D. Peers (ed.) *Warfare and Empires: Contact and Conflict between European and Non-European Military and Maritime Forces and Cultures* (Aldershot, 1997).
A. Starkey, *War in the Age of Enlightenment, 1700–1789* (Westport, Conn., 2003).
F. Tallett, *War and Society in Early Modern Europe, 1495–1715* (London, 1992).
P. H. Wilson, *German Armies: War and German Politics, 1648–1806* (London, 1998).
—— *From Reich to Revolution: German History, 1558–1806* (Basingstoke, 2004).

2 CONFLICT BETWEEN WESTERNERS AND NON-WESTERNERS

A. S. Donnelly, *The Russian Conquest of Bashkiria, 1552–1740* (New Haven, Conn., 1968).
J. Hemming, *Red Gold: The Conquest of the Brazilian Indians, 1500–1760* (2nd edn, London, 1995).
B. Lenman, *Britain's Colonial Wars, 1688–1783* (Harlow, 2001).
J. Lynn (ed.) *Tools of War: Instruments, Ideas and Institutions of Warfare, 1445–1871* (Urbana, Ill., 1990).
W. H. McNeill, *Europe's Steppe Frontier, 1500–1800* (Chicago, Ill., 1964).
G. Parker, *The Military Revolution: Military Innovation and the Rise of the West, 1500–1800* (2nd edn, Cambridge, 1996).
D. B. Ralston, *Importing the European Army: The Introduction of European Military Techniques and Institutions into the Extra-European World, 1600–1914* (Chicago, Ill., 1990).
D. Weber, *The Spanish Frontier in North America* (New Haven, Conn., 1992).

3 THE NATURE OF CONFLICT

C. Duffy, *The Military Experience in the Age of Reason* (2nd edn, London, 1998).
—— *The Fortress in the Age of Vauban and Frederick the Great* (London, 1985).
B. P. Hughes, *Firepower: Weapons' Effectiveness on the Battlefield 1630–1850* (London, 1974).

SELECT BIBLIOGRAPHY

4 WARFARE, 1660–88

T. Barker, *Double Eagle and the Crescent: Vienna's Second Turkish Siege* (Albany, NY, 1967).
M. Glozier, *Marshal Schomberg 1615–1690: 'The Ablest Soldier of His Age': International Soldiering and the Formation of State Armies in Seventeenth-Century Europe* (Brighton, 2005).
J. R. Jones, *The Anglo-Dutch Wars of the Seventeenth Century* (Harlow, 1996).
J. Lynn, *Giant of the Grand Siècle. The French Army: 1610–1715* (Cambridge, 1997).
—— *The Wars of Louis XIV* (Harlow, 1999).
G. Rowlands, *The Dynastic State and the Army under Louis XIV: Royal Service and Private Interest, 1661–1701* (Cambridge, 2002).

5 WARFARE, 1689–1721

D. Chandler, *The Art of Warfare in the Age of Marlborough* (New York, 1976).
R. I. Frost, *The Northern Wars, 1558–1721* (Harlow, 2000).
R. M. Hatton, *Charles XII* (London, 1968).
P. Lenihan, *1690: The Battle of the Boyne* (Stroud, 2003).
J. Pritchard, *In Search of Empire: The French in the Americas, 1670–1730* (Cambridge, 2004).
C. Storrs, *War, Diplomacy and the Rise of Savoy, 1690–1720* (Cambridge, 1999).

6 WARFARE, 1722–55

M. S. Anderson, *The War of the Austrian Succession, 1740–1748* (Harlow, 1995).
C. Duffy, *The Army of Frederick the Great* (London, 1974).
—— *Frederick the Great: A Military Life* (London, 1988).
D. E. Showalter, *The Wars of Frederick the Great* (Harlow, 1996).

7 WARFARE, 1756–74

C. Duffy, *Instrument of War: The Austrian Army in the Seven Years War* (Chicago, Ill., 2000).
L. Kennett, *The French Armies in the Seven Years' War* (Durham, NC, 1967).
J. R. McNeill, *Atlantic Empires of France and Spain: Havana and Louisbourg, 1700–1763* (Chapel Hill, NC, 1985).

8 WARFARE, 1775–91

J. M. Black, *War for America* (Stroud, 1991).
H. Ward, *The War for Independence and the Transformation of American Society* (London, 1999).

9 WARFARE IN THE FRENCH REVOLUTIONARY AND NAPOLEONIC ERA

J. P. Bertaud, *The Army of the French Revolution* (Cambridge, 1988).
T. C. W. Blanning, *The French Revolutionary Wars, 1787–1802* (London, 1996).
G. C. Bond, *The Grand Expedition: The British Invasion of Holland in 1809* (Athens, Ga., 1979).
D. Chandler, *The Campaigns of Napoleon* (London, 1966).
R. Cobb, *The People's Armies* (New Haven, Conn., 1987).
O. Connelly, *Blundering to Glory: Napoleon's Military Campaigns* (Wilmington, Del., 1987).
C. Duffy, *Borodino: Napoleon against Russia* (London, 1972).
—— *Austerlitz* (London, 1977).
J. R. Elting, *Swords around a Throne: Napoleon's Grande Armée* (New York, 1988).
C. Esdaile, *The Spanish Army in the Peninsular War* (Manchester, 1988).
—— *The Wars of Napoleon* (Harlow, 1995).
P. Longworth, *The Art of Victory . . . Suvorov* (New York, 1965).

SELECT BIBLIOGRAPHY

J. Lynn, *The Bayonets of the Republic: Motivation and Tactics in the Army of Revolutionary France* (Urbana, Ill., 1984).
P. Paret, *Yorck and the Era of Prussian Reform* (Princeton, NJ, 1966).
G. E. Rothenberg, *The Art of Warfare in the Age of Napoleon* (Bloomington, Ind., 1978).
F. C. Schneid, *Napoleon's Conquest of Europe: The War of the Third Coalition* (Westport, Conn., 2005).
S. F. Scott, *The Response of the Royal Army to the French Revolution: The Role and Development of the Line Army during 1789–93* (Oxford, 1978).
P. Wetzler, *War and Subsistence: The Sambre-et-Meuse army in 1794* (Las Vegas, Nev., 1985).
C. White, *The Enlightened Soldier: Scharnhorst and the Militärische Gesellschaft in Berlin, 1801–1805* (New York, 1989).

10 NAVAL POWER

J. R. Bruijn, *The Dutch Navy of the Seventeenth and Eighteenth Centuries* (Columbia, SC, 1993).
J. Bromley, *Corsairs and Navies, 1660–1760* (London, 1987).
D. Crewe, *Yellow Jack and the Worm: British Naval Administration in the West Indies, 1739–1748* (Liverpool, 1993).
J. Glete, *Navies and Nations: Warships, Navies, and State-Building in Europe and America, 1500–1860* (Stockholm, 1993).
R. Harding, *The Evolution of the Sailing Navy, 1509–1815* (Basingstoke, 1995).
—— *Seapower and Naval Warfare 1650–1830* (London, 1999).
D. J. Starkey, E. S. van Eyck van Hesling and J. A. de Moor (eds) *Pirates and Privateers: New Perspectives on the War on Trade in the Eighteenth and Nineteenth Centuries* (Exeter, 1997).

11 SOCIAL AND POLITICAL CONTEXTS

F. Anderson, *A People's Army: Massachusetts Soldiers and Society in the Seven Years' War* (Chapel Hill, NC, 1984).
C. Archer, *The Army in Bourbon Mexico, 1760–1810* (Albuquerque, N. Mex., 1977).
J. Black, *Kings, Nobles and Commoners: States and Societies in Early Modern Europe. A Revisionist Introduction* (London, 2004).
J. Canning (ed.) *Power, Violence, and Mass Death in Pre-Modern and Modern Times* (Burlington, Vt., 2004).
J. Chagniot, *Guerre et société à l'époque moderne* (Paris, 2001).
A. Forrest, *Conscripts and Deserters: The Army and French Society during the Revolution and Empire* (Oxford, 1989).
C. Friedrich, *Urban Society in an Age of War: Nördlingen, 1580–1720* (Princeton, NJ, 1979).
M. Gutmann, *War and Rural Life in the Early Modern Low Countries* (Princeton, NJ, 1980).
J. Keep, *Soldiers of the Tsar: Army and Society in Russia, 1462–1874* (Oxford, 1985).
F. Redlich, *De Praeda Militari: Looting and Booty, 1500–1800* (Wiesbaden, 1956).
O. Rian, 'State and Society in Seventeenth-Century Norway', *Scandinavian Journal of History,* 10 (1985), pp. 337–64.
G. E. Rothenberg, B. K. Király and R. F. Sugar (eds) *East European Society and War in the Pre-Revolutionary Eighteenth Century* (Boulder, Col., 1982).
J. Ruff, *Violence in Early Modern Europe 1500–1800* (Cambridge, 2001).
C. B. Stevens, *Soldiers on the Steppe: Army Reform and Social Change in Early Modern Russia* (DeKalb, Ill., 1995).
P. Wilson, 'German Women and War, 1500–1800,' *War in History*, 3 (1996), pp. 127–60.
P. H. Wilson, *War, State, and Society in Württemberg, 1677–1793* (Cambridge, 1995).
—— *Absolutism in Central Europe* (London, 2000).

INDEX

absolutism 165
Africa 14, 24–5
Alaska 19
Albazin 19, 64
Algiers 24, 64, 109, 145
Almanza, battle of 74
Alsace 46, 73, 87
America, North 12, 14, 19–23, 25, 28, 76–7, 93, 97–9, 168
America, South 22–4, 132–3, 168, 173
American War of Independence 105–9, 151–6, 193–4
Amsterdam 9, 114–15, 120
ancien régime x, 1–3, 124, 185
Angola 14, 24
Argaom, battle of 18
aristocracy 111–12, 176–8
army size 53–4, 171
artillery 34, 90, 93, 110, 122–3, 129, 140, 198
Aspern-Essling, battle of 129, 134
Assaye, battle of 18
Auerstädt, battle of 128, 130–1
Austerlitz, battle of 130–1
Australia 14, 22–3, 27, 149, 202
Austria 9–10, 54, 81–3, 89–91, 112, 118, 124, 137–40, 172, 183, 188, 197–8, 202
Austrian Succession, War of the 46, 50, 78, 85–9, 92, 177
Azov 18, 67–8

Balkans 9
Barcelona 74, 190–1

barracks 187
Bavaria 11, 38, 54, 86, 91, 130–1, 134, 138, 141, 168, 172, 186, 190
Bavarian Succession, War of the 112–14
bayonets 2, 18, 31–3, 135
Belgrade 65–7, 85, 116, 202
Belgrade, battle of 35
Bergen-op-Zoom 35, 89, 193
Berthier, Louis-Alexandre 110–11, 128
Bitonto, battle of 83
Blenheim, battle of 11, 33, 73
Blockade 160–1
'blue water' policies 9
Bohemia 87, 89, 94, 113, 167, 183
Borodino, battle of 73, 129, 136
Boyne, battle of the 174
Brazil 22–3
bridges 42
Britain 8, 71, 178, 183, 197
Brittany 167
Brunswick, Charles, Duke of 114–15, 120, 122
Buenos Aires 160
Bunker Hill, battle of 105
Burma 12, 23, 142, 146–7, 179
Bussaco, battle of 135
Buxar, battle of 17

Cairo 131
Calabria 131
California 22
Camden, battle of 106

223

INDEX

Canada 19–20
Cape Breton Island 5, 76, 193
Cape Passaro, battle of 76, 150
Cape Town 152
Carnot, Lazare 124
Caspian Sea 11, 18
Catherine II, the Great 4, 110, 115–16, 150, 174–5
Caucasus 19
cavalry 15, 33–4, 48, 129
Cesmé, battle of 71, 102, 144
Charles Emmanuel III 88, 91, 104
Charles of Lorraine 86–7, 89–90, 95
Charles VI, Emperor 9, 85
Charles XII 65, 69–70, 94, 174, 176
China 13–14, 19, 48, 63, 65, 67, 71, 93, 142–3, 145–7, 158, 193, 202
Chouans 125
Christianity 171
Clausewitz, Carl von 6
Clive, Robert, 15–17
columns 120–2
conscription 111, 128, 179–86
Continental Blockade 132
Cook, James 144
copper sheathing 151–2
'cordon system' 121
Cornwallis, Charles, Earl 16
corps system 128, 135, 139
Corsica 11, 103–4, 123, 190
Courland 189
Crete 14, 61
Crimea 63, 84, 116
Culloden, battle of 88, 191
Cumberland, William, Duke of 9, 87–9, 94, 191–2

Dardanelles 18, 102, 160
Denmark 59, 115, 117, 150, 160, 170, 173, 181, 186, 193
Derbent 18, 81, 142
desertion 181, 185–6
Dettingen, battle of 37, 87, 176
discipline 188, 189

disease 44, 108, 168, 189
divisions 111, 128
Dresden, battle of 138
du Teil, Jean 110
Dumouriez, Charles 122–3
Dundas, David 194
Dupleix, Joseph-François 17
Dutch War of 1672–8 4, 55–6, 58–60

education, military 200
Egypt 12, 27, 133, 142, 159–61, 166
England 52, 59
entrepreneurs, military 112
Estonia 11
Eugene, Prince 2, 75, 83, 90
Eylau, battle of 129, 131

Fehrbellin, battle of 60
Finland 11, 78, 85, 117
Fleurus, battle of 71, 120
Florida 20–1, 76
Folard, Jean-Charles, Chevalier de 79–80, 111
Fontenoy, battle of 35, 37, 87–8, 174
food 187
fortifications 31, 87
France 55–9, 71–6, 118–27, 149, 154, 158, 166, 172, 177–8, 180, 184–6, 188, 197–8, 188–9, 197
Frederick II, the Great xi, 10, 33–6, 78, 85–6, 92, 94–7, 109–10, 112–14, 129, 172, 175–7, 180, 188, 198
French Revolution 30
French Revolutionary War 118–27
Friedland, battle of 131
Fulton, Robert 162–3

Gaelic warfare 34
Geneva 115
Genoa 11, 60, 91, 104, 141, 190, 193
George II 87, 95, 172, 176
George III 12, 53, 155, 157, 160, 167, 174–7, 194
Germantown, battle of 107
Germany 50

224

INDEX

Gibbon, Edward 15
gloire 7–8, 51
Glorious First of June, battle of the 159
Greece 192
Gribeauval, Jean de 110
guerrilla warfare 189–93
Guibert, Jacques, Count 6, 109, 111, 198
Guilford Court House, battle of 106
Gustavus III of Sweden 115, 117

Haidar Ali 12, 17
Haiti 22, 133, 168, 174, 199
Hanover 54, 81–2, 131, 172
Hawaii 143–4
Hesse-Cassel 182
Hohenfriedberg, battle of 87
Hohenlinden, battle of 130
Hume, David 114
Hungary 22, 44, 62, 75, 90–1, 115, 172, 175, 183, 186, 190, 198

India 5, 12, 15–17, 25–7, 29, 48, 59, 64–5, 78, 81, 93, 98–100, 117, 133, 144, 200–1, 203
Industrial Revolution 162
Ireland 132, 161, 199–200
Iroquois 19
Italy 9, 50, 54, 76, 141
Izmail 5, 35, 102, 116

Jacobites 75–6, 88, 150, 179, 190–2
Japan 26, 145–7, 158, 164
Java 79, 162
Jefferson, Thomas 148
Jemappes, battle of 120, 122, 126
Jena, battle of 123, 130–1
Jomini, Antoine-Henri de 6
Joseph II 115–16, 122, 150, 173, 183
Joseph II, Emperor 10

Kagul, battle of 102
Karnal, battle of 25
King Philip's War 64
Kolin, battle of 91, 94, 96–7
Korea 145–6, 158
Kozludji, battle of 102–3

Kunersdorf, battle of 96–7
Lawfeldt, battle of 35, 89, 174, 176
Leipzig, battle of 138
Leopold II, Emperor 10
Leuthen, battle of 36, 94–6
levée en masse 119, 125, 132, 139, 185
Liechtenstein, Prince Joseph 90, 100
Liège 115, 119, 167, 193
Livonia 11
Lloyd, Henry 109
logistics 10–11, 38–40, 42–3, 45–6, 95, 198
Lombardy 39, 42–3, 45, 82–8
Louis XIV 2, 7, 19–20, 31, 46, 49, 51, 52–3, 55–7, 60, 63, 66, 71, 75, 83, 140, 167, 174, 175, 188
Louis XV 174, 177–8
Louisbourg 5
Louisiana 5, 21, 28, 133, 167
Lützen, battle of 123, 129, 137

Maastricht 51, 57, 89
Madrid 167
Malplaquet, battle of 73, 88
maps 198
Marathas 15, 17–18, 25–6, 29, 65, 79, 108, 147
Marengo, battle of 130
Maria Theresa 85–6, 91, 94, 174–5
Marlborough, John, Duke of 2, 10, 73, 75
Marseille 52, 125, 167–8
Mediterranean 202
Mesnil-Durand, François-Jean, Baron de 111
Mexico 200
militias 169–70, 178, 184
Miller, Franz 80
Minden, battle of 33, 37, 111
Minorca 74, 98, 109
Moldavia 192
Mollwitz, battle of 86, 176
Mongolia 67
Mons 71, 193
Montalembert, Marquis de 31
Montenegro 192

INDEX

Morea 62, 66, 147
Morocco 24, 143
Mysore 12, 17, 25, 108, 147, 159

Nadir Shah 12, 19, 25, 81, 83–4, 147
Namur 71, 72, 174
Naples 50, 83, 131
Napoleon 8, 10, 18, 33, 81, 87, 89, 104, 111, 123–4, 126–41, 157, 159–60, 176, 178, 185–6, 200
Napoleonic Wars 127–41
Narva, battle of 69
naval bases 54, 57, 149, 153
Nelson, Horatio 129, 135, 158
Nepal 27, 147
New Zealand 23, 143
Nile, battle of the 159
Nine Years' War 71–2
Norway 117, 132, 160, 167, 186
Nova Scotia 20, 76, 193

oblique order 10
Ochakov 84, 116
Oran 14, 24
Oudenaarde, battle of 73

Paris 169
Patna, battle of 17
Pavian Revolt 126–7
Pensacola 76, 109
Persia 12, 14, 18–19, 81, 83, 133, 142, 146–7
Peru 23
Peter the Great 2, 4, 9, 11, 18, 43, 65, 67–70, 81, 83–4, 150, 168, 172, 175, 181, 202
Philip V of Spain 74
Piedmont 5, 7
Plassey, battle of 16–17, 78
Poland 11, 14–15, 38, 61, 83, 101–2, 115, 118, 126, 178, 190, 196, 199
Polish Succession, War of the 46, 56, 78, 82–3, 151, 177, 196
Pollilore, battle of 17
Poltava, battle of 69–70, 174

Pondicherry 5, 133, 144–5, 193
Pontiac's War 21, 36, 99
Portugal 35, 39–41, 46, 52, 131, 173–4, 186
Potemkin, Prince Gregory 116
Prague 11, 35, 86–7, 94, 113
Prague, battle of 36
Prestonpans, battle of 88
prisoners 166
privateering 6
Prussia 51, 54, 60, 82, 85–6, 89, 92, 94–7, 109, 126, 130–1, 135, 137–40, 172, 181–2, 188, 196–7
Pruth, battle of the 202

Quadruple Alliance, War of the 76, 151
Quebec 11, 76, 98–9, 107, 109, 153, 176
Quiberon Bay, battle of 150

Ramillies, battle of 73
ramrods 32, 45
roads 42–3
Roberts, Michael x, 39
Rossbach, battle of 33, 78, 94–5, 111, 188
Roucoux, battle of 33, 35, 89
Rumford, Count 168, 188
Rumyantsev, Count Peter 101–3
Russia 9, 11, 14, 28, 54, 67–71, 84, 91, 94–7, 115, 123, 127, 130, 133, 136–40, 149–50, 158, 168, 172, 177, 181, 196–7, 199, 202
Russo-Turkish War of 1768–74 100–3

Saint-Domingue 22, 184, 200
Saintes, battle of the 108, 150, 152
Salamanca, battle of 135
Saratoga, battle of 106
Sardinia, kingdom of 7
Savoy-Piedmont 170
Saxe, Maurice, Marshal, Count of 35, 78, 80, 89, 176
Saxony 86, 181
Schaumburg-Lippe, Wilhelm, Count of 40, 109
science 26, 79–80
Scotland 39, 88, 104, 168, 190–2, 194

226

INDEX

Sedgemoor, battle of 190
Selim III 201–3
Serbia 84, 116–17, 132, 199
Seven Years' War 10, 35–7, 45, 54, 78, 91–2, 93–9, 109, 113
Siam 12, 64, 142, 146, 179
Siberia 19, 21, 70
Sicily 4, 50, 59, 83, 167, 202
sieges 5, 122–3
Silesia 85–7, 110, 172, 198
Silistria 5
'small wars' 48–50, 109
Smith, Adam 15
Sobieski, John 61–2, 174
Spain 9, 38, 49–50, 52–3, 76, 81–2, 108, 131–2, 149, 154, 158–60, 167, 173, 166, 173, 183, 189–90, 193, 202
Sri Lanka 27–8, 99–100, 142, 155
St. Gotthard, battle of 61
steamships 145, 162
Stettin 46
Strasbourg 60
Submarines 154, 158, 162–3
subsidies 74
Suvorov, Count Alexander 101–3, 116, 127
Sweden 8, 18, 52, 59, 69, 85, 115, 117, 132, 136, 149, 150, 177, 178, 180, 182–3
Switzerland 180, 190

Taiwan 146–7, 202
Talavera, battle of 135
Tangier 14
Tatars, Crimean 18, 62, 84
Third Battle of Panipat 25, 81
Tibet 65
Tipu Sultan 12, 17, 159
Toulon 73, 123, 125, 149, 155–6, 159–60, 162, 188
Trafalgar, battle of 153, 157, 159
training 187, 201
Transylvania 85, 167
Trenton, battle of 107
Turin, battle of 69, 74, 174

Turkey 18, 26, 28, 61–5, 78, 81, 84–5, 115–17, 130, 132, 136–7, 143, 160, 166, 173, 179, 193, 201–3
Turkey, Ottoman Empire 8, 10, 14
Tyrol 134, 186, 190

Ukraine 52, 61, 69, 172, 192, 197
Ulm, battle of 130
uniforms 173, 175, 188
United Provinces (see also Dutch, Holland) 8–9, 55, 60, 77, 79, 114–15, 149, 159–60, 172, 178, 199
USA 133, 142, 148, 178

Valmy, battle of 120, 122
Vauban, Sébastien Le Prestre de 31, 57, 72, 88
Vegetius, Flavius 80
Vendée 125, 127
Venice 62–3, 66, 168
veterans 188
Victor Amadeus II, 71, 74, 174
Vienna 11, 14, 35, 48, 62, 69, 86, 174
Vimeiro, battle of 135

Wagram, battle of 129, 134
Walcheren 160
Wallenstein, Albrecht von 112
War of 1812 157–8, 162–3
War of Devolution 56
Washington, George 97, 107, 110, 154, 174
Waterloo, battle of 124, 140
weapons 44–5
Wellington, Duke of 18, 135, 140
West Indies 12, 22–3, 28
William III of Orange 9, 59, 63
winter 46

Xinjiang 65, 93

Yorktown, battle of 108–9, 153, 156

Zálákemén, battle of 65
Zeitgeist 6
Zenta, battle of 35, 66
Zorndorf, battle of 96

Introduction to Global Military History
1775 to the Present Day
Jeremy Black

'A lucid and succinct account of military developments around the modern world that combines a truly global coverage of events with thought-provoking analysis. By juxtaposing the familiar with the previously neglected or largely unknown, Jeremy Black forces the reader to reassess the standard grand narrative of military history that rests on assumptions of western cultural and technological superiority ... It should have a wide market on world history courses that are increasingly common parts of American, British and Australian university programmes.' Professor Peter H. Wilson, *University of Sunderland*

'Jeremy Black does an admirable job in distilling a tremendous amount of information and making it comprehensible for students.' Professor Lawrence Sondhaus, *University of Indianapolis*

'An excellent book. Too often, in military studies and histories, the land, air, and maritime aspects are dealt with in separate books. This work integrates all aspects of conflict in a reasonable manner.' Stanley Carpenter, *Professor of Strategy and Policy, US Naval War College, Newport, Rhode Island*

Hb ISBN13: 978-0-415-35394-6   Pb ISBN13: 978-0-415-35395-3
Hb ISBN10: 0-415-35394-7   Pb ISBN10: 0-415-35395-5

Rethinking Military History
Jeremy Black

'Jeremy Black has exercised his formidable powers of historical dissection, critical analysis, and creative cogitation to produce an exciting book…it should spark constructive debate about how historians may better practise their craft.' Theodore F. Cook, *William Paterson University of New Jersey*

'Jeremy Black provides timely arguments against a narrowly technological perception of military history, shaped by Western experience. His survey of five centuries of global warfare shows the shortcomings of this perspective and the necessity to understand the political and cultural aspects of warfare.' Jan Glete, *Stockholm University*

'Formidable.' Paul A. Fideler, *Lesley University*

This must-read study demonstrates the limitations of current approaches, including common generalisations, omissions, and over-simplifications. Engaging theoretical discussions, with reference to specific conflicts, suggest how these limitations can be remedied and adapted, whilst incorporating contributions from other disciplines. Additional chapters provide a valuable and concise survey of the main themes in the study of military history from 1500 to the present day.

Hb ISBN13:978-0-415-27533-0   Pb ISBN 13: 978-0-415-27534-7
Hb ISBN10: 0-415-27533-4   Pb ISBN10: 0-415-27534-2

Available at all good bookshops
For ordering and further information please visit:
www.routledge.com